Advances in
Carbohydrate Chemistry and Biochemistry

Volume 70

Advances in Carbohydrate Chemistry and Biochemistry

Editor
DEREK HORTON
Ohio State University, Columbus, Ohio
American University, Washington, DC

Board of Advisors

DAVID C. BAKER
DAVID R. BUNDLE
STEPHEN HANESSIAN
JÉSUS JIMÉNEZ-BARBERO
YURIY A. KNIREL

TODD L. LOWARY
SERGE PÉREZ
PETER H. SEEBERGER
ARNOLD E. STÜTZ
J.F.G. VLIEGENTHART

Volume 70

AMSTERDAM • BOSTON • HEIDELBERG • LONDON
NEW YORK • OXFORD • PARIS • SAN DIEGO
SAN FRANCISCO • SINGAPORE • SYDNEY • TOKYO
Academic Press is an imprint of Elsevier

Academic Press is an imprint of Elsevier
The Boulevard, Langford Lane, Kidlington, Oxford, OX5 1GB, UK
32, Jamestown Road, London NW1 7BY, UK
Radarweg 29, PO Box 211, 1000 AE Amsterdam, The Netherlands
225 Wyman Street, Waltham, MA 02451, USA
525 B Street, Suite 1800, San Diego, CA 92101-4495, USA

First edition 2013

Copyright © 2013 Elsevier Inc. All rights reserved

No part of this publication may be reproduced, stored in a retrieval system or transmitted in any form or by any means electronic, mechanical, photocopying, recording or otherwise without the prior written permission of the publisher

Permissions may be sought directly from Elsevier's Science & Technology Rights Department in Oxford, UK: phone (+44) (0) 1865 843830; fax (+44) (0) 1865 853333; email: permissions @elsevier.com. Alternatively you can submit your request online by visiting the Elsevier web site at http://elsevier.com/locate/permissions, and selecting *Obtaining permission to use Elsevier material*

Notice
No responsibility is assumed by the publisher for any injury and/or damage to persons or property as a matter of products liability, negligence or otherwise, or from any use or operation of any methods, products, instructions or ideas contained in the material herein. Because of rapid advances in the medical sciences, in particular, independent verification of diagnoses and drug dosages should be made

ISBN: 978-0-12-408092-8
ISSN: 0065-2318

British Library Cataloguing in Publication Data
A catalogue record for this book is available from the British Library

Library of Congress Cataloging-in-Publication Data
A catalog record for this book is available from the Library of Congress

For information on all Academic Press publications
visit our website at store.elsevier.com

Printed and bound in USA

13 14 15 16   11 10 9 8 7 6 5 4 3 2 1

CONTENTS

CONTRIBUTORS vii

PREFACE ix

STRUTHER ARNOTT 1934–2013
CHANDRASEKARAN RENGASWAMI

Seven Decades of "*Advances*"
DEREK HORTON

I. Introduction and Background 13
   1. The Launching of *Advances* and Its Policies 13
   2. Carbohydrate Nomenclature and Indexing 14
   3. Carbohydrate Reference Books 16
   4. Founding of the Journal *Carbohydrate Research* 17
II. Editors' Prefaces for Volumes 1–70 of *Advances* 18
III. Concluding Remarks 207
    References 208

AUTHOR INDEX 211
SUBJECT INDEX 221

CONTRIBUTORS

Derek Horton, Department of Chemistry, Ohio State University, Columbus, Ohio, USA, and Department of Chemistry, American University, Washington, DC, USA

Chandrasekaran Rengaswami, Whistler Center for Carbohydrate Research, Department of Food Science, 745 Agriculture Mall Drive, Purdue University, West Lafayette, IN 47907-2009, USA

PREFACE

This 70th volume of *Advances* constitutes a small supplementary issue that pauses to review the evolution of the series since it began under the title *Advances in Carbohydrate Chemistry* toward the close of the Second World War, and looks toward its future as *Advances in Carbohydrate Chemistry and Biochemistry* in the ongoing documentation of a vastly expanded field of scientific disciplines where carbohydrates play a role.

The volume opens with an appreciation of the career of Struther Arnott and his contributions to our understanding of polysaccharide structure by use of X-ray diffraction studies of oriented fibers. The celebrated work of Crick and Watson, unraveling the structure and anticipating the all-encompassing biological role of DNA, was based on the X-ray diffraction studies by Maurice Wilkins and Rosalind Franklin at King's College, London. Arnott's later work in Wilkins' laboratory greatly extended our knowledge and understanding of the three-dimensional architecture of a wide range of polynucleotide chains by application of his Linked-Atom Least-Squares (LALS) methodology. Arnott's biographer, Rengaswami Chandrasekaran, details his subsequent researches at Purdue University in the USA that extend the compass of the technique to the structural characterization of a broad range of polysaccharides having relevance in both biological and technological areas. Equipped with great research talent, coupled with high administrative capacity, Struther Arnott fulfilled important leadership roles at Purdue and subsequently at St. Andrews University in his native Scotland. The Bibliography appended to the memoir provides a comprehensive source of reference to the many polysaccharide structures expertly documented by the Arnott research teams.

The *Advances* series set out in 1944 with the objective of presenting critical and integrating reports, understandable by the general reader as well as the specialist, on a wide range of topics having carbohydrates as a common theme. The report "Seven Decades of Advances" in this current volume assesses broadly the literature record on carbohydrates, as documented in over 350 articles published in this series during the seventy years of its existence, and its relation to research papers published in primary journals, as well as information in reference books, monographs, and text books that constitute the secondary literature.

At the time this series began, reports on original research were being published in a wide range of national journals and in many different languages. Some papers described sound, original, and novel work, but others may have had errors of fact, be of dubious originality, or had failed to give credit to relevant prior work. Although the more-prestigious national journals exercised a level of quality control, the peer-review system had not then developed to its current broad extent. In areas such as the carbohydrates, where a uniform system of nomenclature did not exist, there was a great deal of confusion in naming compounds and in their structural depiction.

It was the goal of a small group of leading carbohydrate chemists, notably Melville L. Wolfrom and Claude S. Hudson in the USA, and Sir Norman Haworth in Great Britain, and with the strong support of a far-sighted publisher, Kurt Jacoby (who had just founded Academic Press), to produce an annual book series featuring reports from invited experts on a wide distribution of topics related to carbohydrates. These areas would include simple sugars, glycosides, oligosaccharides, and polysaccharides, and deal with their structures, chemical and biochemical reactions, analytical chemistry, food and fiber technology, and other aspects.

Their endeavors set the stage from which the 70 volumes of *Advances* subsequently developed during a span of seven decades, averaging one volume per year. The authors contributing to the early volumes came mainly from North America and Great Britain, but the sources have expanded progressively to include articles from authors in countries throughout the world, from Europe, Australia, New Zealand, South America, Asia, and Africa. During this period, Academic Press took a prominent position in documenting other aspects of the carbohydrate field, with such series as Whistler's *Methods* and Aspinall's *Polysaccharides* volumes, and the Pigman books. When Academic Press became part of the Elsevier line, the breadth was further enhanced with the founding of the journal *Carbohydrate Research*.

The articles published in *Advances* have adhered to the original policies of its founders, and many of them remain definitive treatments of their subject, while others reflect advances in areas still in the course of rapid development. Collected here in chronological sequence are the published Prefaces to the individual volumes, and these describe the subject material in all of the individual articles, as well as providing contemporary commentary at the time the volumes were published.

Parallel with the evolution of the series there has been important progress in defining the language of scientific communication through agreements on standardized nomenclature; this has been regularly documented in the pages of the *Advances* volumes. It is a tribute to the cooperative efforts volunteered by a number of individuals that the standardization reflected in the volumes has served as a benchmark for widespread adoption of these nomenclature recommendations by authors throughout the world.

An important feature of *Advances* has been regular obituary reports that document the lives and motivations of major figures who have made significant contributions to the carbohydrate field. For the particular benefit of researchers who may not have had the opportunity to meet or hear these past leaders in person, this volume features a collection of portraits of all of those past leaders whose stories have been told in the various volumes.

The editorship of the series was, with one brief interruption, in the hands of its prime motivator, M. L. Wolfrom, until his death in 1969. Robert Stuart Tipson shared the editorship with Wolfrom, starting with Volume 8 in 1954. He worked jointly with the present editor until Volume 48 (1990), with the latter continuing thereafter.

The initiative of Academic Press to make back issues of *Advances* accessible electronically was significantly enhanced when the Elsevier organization established electronic access to a wide swath of the literature on carbohydrates through Science Direct, Scopus, and other sources. The massive chemical database of *Chemical Abstracts* provides records on individual carbohydrates and their derivatives, estimated to comprise at least 5% of the 73 million or more substances in the CAS Registry, and all are accessible electronically through SciFinder.

A bibliometric analysis of *Advances*, made in early 2013 by Professor Todd Lowary, revealed that 68 volumes of the series since the first issue had a total of 358 individual articles, carrying altogether 24,558 literature citations. These reports were cited 20,582 times, with an average of 70 citations per article and a 2012 impact factor of 7.133. The five most-cited reports were Bock and Pedersen's ^{13}C-NMR of monosaccharides article in Volume 41, with 1303 citations, the Schauer report on sialic acids in Volume 40 (858 citations), the Volume 41 ^{1}H-NMR glycoprotein article by Vliegenthart, Dorland, and van Halbeek (843 citations), the Schmidt–Kinzy trichloroacetimidate methodology in Volume 50, cited 694 times, and Legler's article in Volume 48 on the mechanism of glycoside hydrolases, which had 593 citations.

The content of this *Advances* series provides a large and unified source of reliable reference material contributed by noted experts in the field, and is presented in clear, understandable English without unnecessary jargon or incomprehensible acronyms. The excellent support of the contributing authors and members of the Board of Advisors is expected to assure a fruitful ongoing continuation of the series in the service of the carbohydrate community.

DEREK HORTON

Washington, D.C.
Columbus, Ohio, September 2013

STRUTHER ARNOTT

1934–2013

Large was his bounty, and his soul sincere,
Heaven did a recompense as largely send:
He gave to Misery all he had, a tear,
He gained from Heaven ('twas all he wished) a friend.
  "Elegy Written in a Country Churchyard"—Thomas Gray (1751)

The scientific world and the field of fiber crystallography have lost a major figure with the passing of a respected and beloved colleague, Professor Struther Arnott, aged 78, on April 20, 2013, at his residence in Doncaster, a small market town in south Yorkshire, England. Motivated by curiosity, drive, and enthusiasm, Dr. Arnott was internationally acclaimed for his meticulous research projects and scrupulous contributions to the three-dimensional structures of nucleic acids, polypeptides, and polysaccharides. He also took up challenging administrative positions of increasing complexity—Department Head, Dean of Graduate School, Vice President for Research, and University Principal—throughout a rich professional career.

An achiever, admired and appreciated by one and all, Dr. Arnott was born on September 25, 1934, in Larkhall, a Scottish town south of Glasgow in the county of Lanarkshire. He received exemplary high-school education between 1945 and 1952 at the Hamilton Academy, a prestigious school founded in 1588, whose graduates enter the top universities in the country. His scholastic accomplishments were evident even at that young age; he was awarded the school's Cuthbertson Science Scholarship and Lorimer Mathematical Scholarship.

In 1952, at the age of 17, he earned a gold medal for general scholarship, as well as a silver medal for both chemistry and mathematics. In the same year, he secured first place in science in the University of Glasgow Open Bursary Competition. Equipped with this strong educational foundation, he entered the University of Glasgow, where in 1956 he got his B.Sc. degree in Chemistry and Mathematics.

Inspired by a passion for science and set on visualizing the innate arrangement of molecules and their associated functionality at the atomic level, Dr. Arnott worked with Drs. D.H.R. Barton and J.M. Robertson in the Department of Chemistry, University of Glasgow, as a graduate student, mastering the basics of X-ray crystallography. He then undertook a heuristic investigation on limonin ($C_{26}H_{30}O_8$), a six-ring furanolactone that is the bitter principle of citrus fruits, a particularly arduous task during a time when computers and programs were still quite primitive. Despite the large number of atoms in the molecule, he successfully solved the limonin structure by using the phases calculated from the position of the iodine atom in its iodoacetate derivative—a Herculean task indeed. This earned him his Ph.D. degree in 1960.

Dr. Arnott then moved on to King's College, London, to join Dr. M.H.F. Wilkins, whose ambition was to unravel the molecular details of the DNA double-helix proposed in 1953 by Crick and Watson. Dr. Arnott's strong proficiency in mathematics was a major driving force behind his taking up this challenge. He soon developed a suitable and elegant methodology, involving several complex equations in matrix notation, for the refinement of nucleic acid structures. During a talk on "Diffraction studies of biological molecules" given in 1963 at the Aberdeen meeting of the British Association for the Advancement of Science, Dr. Arnott remarked, "To date, in the process of integrating biology with the more physical sciences, X-ray diffraction has played a leading part. It is for this reason that people like myself, knowing little botany and less zoology, have ventured into the fringes of the biological jungle in the hope of obtaining fame and fortune." True to these words, he immersed himself in research at the Medical Research Council Biophysics Unit at King's College, producing several seminal scientific contributions in the arena of X-ray fiber diffraction in general and nucleic acids in particular.

Experimental X-ray observations on single crystals contain an enormous number of reflections, more than adequate to allow determination of the precise atomic coordinates. By contrast, stretched fibers of helical polymers, because of their rotational symmetry, produce a mere handful of reflections, insufficient for a similar analysis. In this context, Dr. Arnott's statement of 1963 is worth noting: "While Perutz and Kendrew had to move mountains of data in their analysis (of hemoglobin and myoglobin), we have had the equal embarrassment of gleaning our structural information from molehills of data." A way out of this misery, he believed, was to recast the problem in terms of fewer variables. By linking each atom in a helical repeating-unit in a contiguous "tree geometry," with fixed bond lengths and bond angles (as known parameters), while conformational angles were allowed to be variable or dependent on one another, the atomic coordinates could be generated by appropriate mathematical formulation. Elucidation of the correct molecular shape, compatible with the sparse X-ray data, would then be within reach by scrupulous refinement of the small number of conformational angles. Using this simple concept, he elegantly developed the Linked-Atom Least-Squares (LALS) methodology (1966) for solving molecular structures. Necessity was the mother of invention! From then on, Dr. Arnott's novel computer program, a major milestone in fiber crystallography, paved the way for analyzing and understanding the architectures of biopolymer systems.

The test beds for validating the new procedure were the alpha-helix (1966, 1967) and beta-pleated sheet (1967) of poly-L-alanine and that of poly-L-proline II (1968). Later on, a refinement performed using rich single-crystal data of (Pro-Pro-Gly)$_{10}$ provided the fine structural features of collagen at high resolution (1981). Noteworthy amongst Dr. Arnott's pioneering successes was the clever application of Fourier methods to confirm and conclude that the Watson–Crick base-pairing scheme was the only viable option matching the X-ray data of B-DNA. This scheme thereby eliminated such alternatives as the Hoogsteen and reverse Hoogsteen schemes known at that time (1965). Dr. Arnott then embarked on double-helical RNA (1967) and refined both structures (1969).

In 1970, the United States became Dr. Arnott's new home when he joined the faculty at Purdue University in West Lafayette, Indiana, as Professor in the Department of Biological Sciences. He set up his fiber-diffraction laboratory, with new X-ray generators and pin-hole cameras fabricated in the departmental machine shop, by meticulously following the sketches brought from King's College. Dr. W.E. Scott, arriving as a postdoctoral associate, was delegated to look after the X-ray generators in addition to his regular research. Dr. Scott had just received his Ph.D., mentored by the well-known British carbohydrate chemist Dr. D. A. Rees of Unilever Research. Dr. Rees had no experience in solving fiber structures using X-ray methods but was anxious to decipher the intriguing structure–function relationships in polysaccharides, especially in carrageenan. Dr. Arnott, while at King's College, had helped Dr. Rees to learn the theory and practice of X-ray diffraction of stretched fibers. Now at Purdue, Dr. Arnott entered the carbohydrate arena enthusiastically. He swiftly solved the structure of the gel-forming polysaccharide iota-carrageenan using his LALS protocol, and he published together with Drs. Scott and Rees its fascinating double-helical arrangement (1974). This successful collaboration not only evidently catalyzed mutual respect and lifelong admiration between these two industrious scientists but also enhanced further scientific exploration in, and contributions to, the polysaccharide field. The Arnott research group gradually grew in size, adding graduate students, postdoctoral research associates, and visiting scholars, each working on different structural aspects of nucleic acids and polysaccharides. While at Purdue, Dr. Arnott mentored eight graduate students and 17 research associates, as well as various visiting scientists. He is credited internationally with a total of 166 research publications.

I had the privilege of joining Dr. Arnott's team in 1971 to pursue nucleic acid research. As days went by, I was increasingly fascinated by his vocabulary, scientific knowledge, and mentoring style. During the subsequent three decades and beyond, our combined efforts led to successful revelation of the intricate atomic arrangements of several native as well as synthetic DNAs of repeating nucleotide sequences—a period that witnessed rich scientific contributions to the field of nucleic acids. I also had the privilege of fulfilling numerous requests from laboratories around the world for the LALS program on magnetic tape; this was before the emergence of electronic file transfers. Even more memorable was the laboratory tradition of celebrating the fruits of hard labor—a bottle of champagne (Korbel Natural) per author of each peer-reviewed manuscript accepted for publication—an eagerly awaited event enjoyed almost every (other) week by one and all in the fiber team. This tradition continued in my group in the years to follow.

It was not uncommon to see Dr. Arnott in his office looking through the microscope, slowly turning the knurled knob of the fiber puller a little at a time, so as to stretch and orient the fiber under controlled humidity. His team was equally diligent in producing suitable specimens for conducting careful structural analyses of industrially important gel-forming polysaccharides, performed to rationalize the molecular basis of gelation. Such materials are routinely exploited to texture foods and for use as biologically compatible materials in medical applications. Results emerging from the Arnott group demonstrated the carrageenan-like double-helical structures for agarose and its derivatives (1974); the cation-mediated interactions among pectin single-helices that stabilize junction zones, explaining why divalent cations produce stronger gels than do monovalent cations (1981); and similar interactions being the origin of gelation in gellan, which adopts a slim and extended double-helical form (1988).

In addition, his attention focused on the medically important glycosaminoglycans of vertebrate connective tissues, such as hyaluronate from synovial fluid and vitreous humor of the eye, and chondroitin and dermatan sulfates from cartilage and skin. Notable was the existence of hyaluronic acid in its nine distinct allomorphs with 3-fold and 4-fold helical structures (1975); under certain circumstances, the latter could relax and assume a lower symmetry (1975); the helix pitch could significantly contract or expand—as would a spring—induced by pH, water, and cations (1975); and a novel antiparallel double-helix stable at extremely low pH (3.0–4.0) in the presence of potassium ions (1975, 1983), in line with the versatile functionality of this polysaccharide. This work embodied the first systematic investigation that revealed the orchestrated roles of atomic interactions for the observed polymorphism. According to the results reported, transition from one form to another was achieved by breaking-old-and-making-new hydrogen bonds in the helix, and by deploying difference Fourier maps, cations were traced to be the key players. Furthermore, similar helical structures and atomic interactions were demonstrated for three other glycosaminoglycans, chondroitin sulfate (1978, 1983), dermatan sulfate (1973, 1983), and keratan sulfate (1974). Based on these findings, the Arnott team visualized how cations could specifically transform the polymorphic polysaccharide structures in the intercellular matrices of animal and human tissues.

Befitting his scientific contributions in this field, Dr. Arnott joined Dr. Rees and his colleague Dr. E.R. Morris as co-editor of the book *Molecular Biophysics of the Extracellular Matrix*, published in 1984. Later on, Dr. Arnott and my team together produced a revised packing arrangement of helices in alginate gels, providing, for the first time, the junction-zone details (2000)—far beyond the scope of the "egg-box" model cartooned 40 years earlier.

From similar rigorous investigations of nucleic acids, the Arnott group consolidated early DNA and RNA structures and then determined the molecular architectures of many new allomorphs. Significant contributions were as follows: Single-helical poly(C) (1976). Double-helical—poly(dG)·poly(dC) (1974); poly(dA)·poly(dT) (1983); poly d(AT)·poly d(AT) (1974); poly d(IC)·poly d(IC) (1983); poly d(GC)·poly d(GC) (1983). Triple-helical—poly(dT)·poly(dA)·poly(dT) (1974, 2000); poly(U)·poly(A)·poly(U) (1973, 2000); poly(I)·poly(A)·poly(I) (1973, 2000). Four-stranded helix of poly(I) (1974). Left-handed Z-DNA—poly d(GC)·poly d(GC) (1980). A novel unwound DNA (1980). A seminal article emerged on the polymorphism of DNA double-helices, highlighting the variations in molecular morphology influenced by nucleotide-base sequence (1980); later, a compendium on the shapes of nucleic acid helices was published (1989). The wide variety of DNA and RNA structures enabled us all to appreciate the extent to which polynucleotide sequences present themselves to interacting proteins and drugs in diverse ways.

During 1975–1980, Dr. Arnott served as Head of the Department of Biological Sciences at Purdue. He encouraged fellow faculty members to investigate more complex biological systems, augmented developmental biology and physiology, and established modern ecology. These administrative chores interfered little with his commitment to research. Although Purdue has no medical school, he successfully persuaded the National Cancer Institute to set up a Cancer Center and the Arthritis Foundation to support his own research on connective tissue.

Armed with admirable administrative skills that paralleled his scientific abilities, Dr. Arnott was selected in 1980 to serve at Purdue as Vice President for Research and

Dean of the Graduate School. He performed the juggling act of tackling dual responsibilities as well as the biology professorship, with finesse through 1985. As Research Vice President, he distributed internal research funds and supervised all research centers and interdisciplinary graduate programs (such as American Studies, Biochemistry, Comparative Literature, Plant Physiology, and Public Policy). As Graduate Dean, he maintained and delivered academic quality control (promotion and tenuring of faculty members, admission and retention of graduate students) throughout the five Purdue campuses (65,000 students).

In 1986, Dr. Arnott returned to his native Scotland as Principal and Vice Chancellor (namely, President and Chief Executive) of The University of St. Andrews, Scotland's first university, founded in 1413. As stated by him, "My task was to maintain ancient distinction and enhance modern impactfulness, despite a rural setting and a small (5500 student) size. The outcome has been that Edinburgh (a 15,000-student civic university founded in 1585) and St. Andrews now share the top two positions for research in Scotland, as ranked by the UK-wide 1996 Research Assessment Exercise. St. Andrews and Edinburgh also share the top two teaching positions, as ranked by the continuing Teaching Quality Assessment in Scotland. Moreover, for each year in the UK, St. Andrews has invariably been near the top of any ranking of universities that measures the employability of its graduates. According to the (London) Times' independent assessment of UK universities for its 1997 Good Universities' Guide, St. Andrews was hailed as 'Scotland's Finest'. These comparative advantages have been achieved by transparent planning, robust finances, novel fund-raising, and investment in up-to-date facilities, while preserving St. Andrews' historic image. Relentless discrimination in recruiting new academic and managerial colleagues has been important, as has been the maintenance of a cosmopolitan population of both senior and junior members. Students come from Europe (10%), the US (10%), Japan (5%), and other parts of the UK (40%) as well as from Scotland (35%) to enjoy a high-quality melting-pot experience in an entirely residential university."

Dr. Arnott's vision of a university received royal endorsement in the year 2000 with the recruitment of Prince William of Wales. At this high point, after a decade and a half of successful scholarship and an administrative career, Dr. Arnott relinquished the helm. In recognition of his stellar contributions, St. Andrews named a lecture hall in his honor.

In 2002, Dr. Arnott accepted a Visiting Professorship at the Imperial College, London, and a similar position in 2003 at the School of Pharmacy of the University of London. In addition to regularly dropping by the two laboratories to "rub shoulders" with colleagues and keep abreast of current research progress in biopolymers, he took time to write two historical articles on nucleic acids (2005, 2006). Furthermore, he redefined the structure for crystalline natural rubber, a subject of controversy for at least 60 years, and published the new and exciting results (2006).

Dr. Arnott's expertise in the structures of not-fully-crystalline materials resulted in consultancies with Unilever (England), DuPont (Delaware, USA), and Proctor and Gamble (Ohio, USA). Because of his contributions to the theoretical and practical aspects of biopolymers, he was chosen as a member of the US National Academy delegation to the Soviet Union, twice, in 1972 and 1975, and a member of the National Science Foundation Commission on Crystallography in 1982. He also served as a member of the Editorial Board of the *Journal of Theoretical Biology* (1976–1978) and the *International Journal of Biological Macromolecules* (1978–1980). Over a span exceeding three decades, his scholarly achievements in molecular physics and biology were recognized by multiple leading institutions. In 1970, the Royal Society of Chemistry, London, named him a Fellow (FRSC). Twice, in 1980 and 1985, Oxford University named him a Senior Fellow. And in 1985, he was also a Memorial Fellow of Guggenheim Foundation; The Royal Society of London honored him as a Fellow (FRS). In 1988, The Royal Society of Edinburgh honored him as a Fellow (FRSE). In 1994, St. Andrews (Laurinburg, North Carolina) honored him with Doctor of Science (Sc.D.). In 1996, the British royalty made him a Commander of the Order of the British Empire (CBE). In 1997, he was a Distinguished Visiting Fellow of the Japan Society for Advancement of Science. In 1998, Purdue University awarded him Doctor of Science (D.Sc.), and a year later, St. Andrews, Doctor of Laws (LLD).

Dr. Arnott's departure from Purdue University in 1986 coincided with the establishment of the Whistler Center for Carbohydrate Research in the Department of Food Science. Two scientists from his group—Dr. R.P. Millane and myself—were hired into faculty positions by the Center's first director Dr. J.N. BeMiller, while Dr. Arnott was retained as an Adjunct Professor of the Center. The fiber-diffraction activities continued in my laboratory and grew with additional focus on delineating the structure–function relationships of polysaccharides and nucleic acids. The following years witnessed the birth of the next generation of fiber diffractionists, the appearance of superfast computers on every desk, and the availability of improved LALS software on laptops. Thrilled and invigorated by such advancements, Dr. Arnott himself continued to tackle certain complex structures from the comfort of his own home. He would often call me for stimulating brainstorming sessions where we strategized on what to do next. My team has been fortunate to have closely interacted with him on many occasions. Thanks to the lessons I learned from him, they were passed on gladly to my group. Fiber diffraction on biomaterials has now risen to the next level and steered toward exploiting stable networks as vehicles for the delivery of bioactive compounds. I view this as my humble tribute to this stalwart fiber diffractionist of contemporary times.

Dr. Arnott was a stern decision-maker who vehemently defended his viewpoints with valid reasons. Blessed with old-fashioned kindness and decency, during official meetings or otherwise, he invariably initiated a conversation in an informal manner, beginning by asking about the family. He was a patient listener and very generous man who never hesitated to help when asked. He had an eye for particulars. He listened to classical music with serious intent. While living in the UK, he and his wife, Greta, enjoyed watching plays of literary value at London theaters. Reading books on history, literature, and politics was one of his favorite pastimes. He loved gardening to ease stress. The Principal, frequently seen out walking—weather permitting—along the Fife sand dunes armed with binoculars, was a devout ornithologist. Gifted with an extraordinary memory, he enjoyed discussing what he had read or seen the last month or even ten years ago with guests at social gatherings. He was eloquent and pleasant while mingling with the crowd. He returned to St. Andrews in early 2013 to deliver a lecture on the history of DNA at Arnott Hall, and it is no surprise that this was delivered magnificently—without notes. The entire audience pounded applause enthusiastically, the same way it had done when he lectured at the Adam Smith Institute on Industry Matters almost 25 years earlier!

Dr. Arnott stimulated colleagues and pupils with enthusiasm for his field and academia in general. With his unfailing impeccable suit and smiling face, he was always a vivid person blessed with a knack for engaging in intelligent conversations with peers, students, and children of all ages. The pleasant get-togethers with friends and colleagues over the years have left indelible impressions and inspirations forever.

Dr. Arnott is survived by his lovely wife, their two sons, and grandchildren. He is sorely missed.

CHANDRASEKARAN RENGASWAMI

BIBLIOGRAPHY

Carbohydrate Articles by Struther Arnott

1. S. Arnott and W. E. Scott, Accurate X-ray diffraction analysis of fibrous polysaccharides containing pyranose rings. Part 1. The linked-atom approach, *J. Chem. Soc. Perkin Trans. II* (1972) 324–335.
2. I. C. M. Dea, R. Moorhouse, D. A. Rees, S. Arnott, J. M. Guss, and E. A. Balazs, Hyaluronic acid: A novel, double helical molecule, *Science*, 179 (1973) 560–562.
3. S. Arnott, J. M. Guss, D. W. L. Hukins, and M. B. Mathews, Mucopolysaccharides: Comparison of chondroitin sulfate conformations with those of related polyanions, *Science*, 180 (1973) 743–745.
4. S. Arnott, J. M. Guss, and D. W. L. Hukins, Dermatan sulfate and chondroitin 6-sulfate conformations, *Biochem. Biophys. Res. Commun.*, 54 (1973) 1377–1383.

5. S. Arnott, D. W. L. Hukins, R. L. Whistler, and C. W. Baker, Molecular conformation in gels of cellulose sulfate, *Carbohydr. Res.*, 35 (1974) 259–263.
6. S. Arnott, J. M. Guss, D. W. L. Hukins, I. C. M. Dea, and D. A. Rees, Conformation of keratan sulfate: A study of stereochemical relationships within the glycosaminoglycan family, *J. Mol. Biol.*, 88 (1974) 175–184.
7. S. Arnott, W. E. Scott, D. A. Rees, and G. G. A. McNab, Iota-carrageenan: Molecular structure and packing of polysaccharide double helices in oriented fibres of divalent cation salts, *J. Mol. Biol.*, 90 (1974) 253–267.
8. S. Arnott, A. Fulmer, W. E. Scott, I. C. M. Dea, R. Moorhouse, and D. A. Rees, The agarose double helix and its function in agarose gel structure, *J. Mol. Biol.*, 90 (1974) 269–284.
9. S. Arnott, J. M. Guss, and W. T. Winter, Glycosaminoglycan conformations, in H. C. Slavkin and R. C. Greulich, (Eds.), *Proceedings of the Second Santa Catalina Island International Symposium on the Influences of the Extracellular Matrix on Gene Expression*, Academic Press, New York, 1975, pp. 399–407.
10. J. M. Guss, D. W. L. Hukins, P. J. C. Smith, W. T. Winter, S. Arnott, R. Moorhouse, and D. A. Rees, Hyaluronic acid: Molecular conformations and interactions in two sodium salts, *J. Mol. Biol.*, 95 (1975) 359–384.
11. W. T. Winter, P. J. C. Smith, and S. Arnott, Hyaluronic acid: Structure of a fully extended 3-fold helical sodium salt and comparison with the less extended 4-fold helical forms, *J. Mol. Biol.*, 99 (1975) 219–235.
12. W. T. Winter, J. J. Cael, P. J. C. Smith, and S. Arnott, Hyaluronic acid conformations and interactions, in J. C. Arthur, Jr., (Ed.), *Cellulose Chemistry and Technology, ACS Symposium Series*, Vol. 48, 1977, pp. 91–104.
13. R. Moorhouse, M. D. Walkinshaw, W. T. Winter, and S. Arnott, Solid state conformations and interactions of some branched microbial polysaccharides, in J. C. Arthur, Jr., (Ed.), *Cellulose Chemistry and Technology, ACS Symposium Series*, Vol. 48, 1977, pp. 133–152.
14. W. T. Winter and S. Arnott, Hyaluronic acid: The role of divalent cations in conformation and packing, *J. Mol. Biol.*, 117 (1977) 761–784.
15. R. Moorhouse, W. T. Winter, S. Arnott, and M. E. Bayer, Conformation and molecular organization in fibers of the capsular polysaccharide from an *Escherichia coli* M41 mutant, *J. Mol. Biol.*, 109 (1977) 373–391.
16. R. Moorhouse, M. D. Walkinshaw, and S. Arnott, Xanthan gum—Molecular conformation and interactions, in P. A. Sandford and A. Laskin, (Eds.), *Extracellular Microbial Polysaccharides, ACS Symposium Series*, Vol. 45, 1977, pp. 90–102.
17. S. Arnott, Ordered conformations of gel-forming polysaccharides obtained by X-ray diffraction analysis of oriented fibres, in G. G. Birch and R. S. Shallenberger, (Eds.), *Developments in Food Carbohydrates*, Applied Science Publ., Ltd., Barking, England, 1977, pp. 43–60.
18. S. Arnott and W. T. Winter, Details of glycosaminoglycan conformations and intermolecular interactions, *Federation Proc.*, 36 (1977) 73–78.
19. W. T. Winter, S. Arnott, D. H. Isaac, and E. D. T. Atkins, Chondroitin 4-sulfate: The structure of a sulfated glycosaminoglycan, *J. Mol. Biol.*, 125 (1978) 1–10.
20. J. J. Cael, W. T. Winter, and S. Arnott, Calcium chondroitin 4-sulfate: Molecular conformation and organization of polysaccharide chains in a proteoglycan, *J. Mol. Biol.*, 125 (1978) 21–42.
21. W. T. Winter and S. Arnott, Secondary and tertiary structure of glycosaminoglycans and proteoglycans, in J. D. Gregory and R. W. Jeanloz, (Eds.), *Glycoconjugate Research: Proceedings of the 4th International Symposium on Glycoconjugates*, Vol. I, Academic Press, New York, 1979, pp. 321–323.
22. K. Okuyama, S. Arnott, R. Moorhouse, M. D. Walkinshaw, E. D. T. Atkins, and C. Wolf-Ullish, Fiber diffraction studies of bacterial polysaccharides, in A. D. French and K. H. Gardner, (Eds.), *Fiber Diffraction Methods, ACS Symposium Series*, Vol. 141, 1980, pp. 411–427.

23. M. D. Walkinshaw and S. Arnott, Conformations and interactions of pectins. I: X-ray diffraction analyses of sodium pectate in neutral and acidified forms, *J. Mol. Biol.*, 153 (1981) 1055–1073.
24. M. D. Walkinshaw and S. Arnott, Conformations and interactions of pectins. II: Packing of pectinic acid and calcium pectate as models for junction zones in gels, *J. Mol. Biol.*, 153 (1981) 1075–1085.
25. S. Arnott, D. A. Rees, and E. R. Morris, (Eds.), *Molecular Biophysics of the Extracellular Matrix*, Humana Press, New Jersey, 1984, pp. 1–189.
26. S. Arnott and A. K. Mitra, X-ray diffraction analyses of glycosaminoglycans, in S. Arnott, D. A. Rees, and E. R. Morris, (Eds.), *Molecular Biophysics of the Extracellular Matrix*, Humana Press, New Jersey, 1984, pp. 41–67.
27. A. K. Mitra, S. Arnott, and J. K. Sheehan, Hyaluronic acid: Molecular conformation and interactions in the tetragonal form of the potassium salt containing extended chains, *J. Mol. Biol.*, 169 (1983) 813–827.
28. A. K. Mitra, S. Raghunathan, J. K. Sheehan, and S. Arnott, Hyaluronic acid: Molecular conformations and interactions in the orthorhombic and tetragonal forms containing sinuous chains, *J. Mol. Biol.*, 169 (1983) 829–859.
29. S. Arnott, A. K. Mitra, and S. Raghunathan, The hyaluronic acid double helix, *J. Mol. Biol.*, 169 (1983) 861–872.
30. A. K. Mitra, S. Arnott, E. D. T. Atkins, and D. H. Isaac, Dermatan sulfate: Molecular conformations and interactions in the condensed state, *J. Mol. Biol.*, 169 (1983) 873–901.
31. R. P. Millane, A. K. Mitra, and S. Arnott, Chondroitin-4-sulfate: Comparison of the structures of the potassium and sodium salts, *J. Mol. Biol.*, 169 (1983) 903–920.
32. A. K. Mitra, R. P. Millane, S. Raghunathan, J. K. Sheehan, and S. Arnott, Comparison of glycosaminoglycan structures induced by different monovalent cations as determined by X-ray fiber diffraction, *J. Macromol. Sci. Phys.*, B24, ((1985) 21–38.
33. R. Chandrasekaran, R. P. Millane, J. K. Walker, S. Arnott, and I. C. M. Dea, The molecular structure of the capsular polysaccharide from Rhizobium trifolii, in V. Crescenzi, I. C. M. Dea, and S. S. Stivala, (Eds.), *Recent Developments in Industrial Polysaccharides*, Gordon and Breach, New York, USA, 1987, pp. 111–118.
34. R. Chandrasekaran, R. P. Millane, S. Arnott, and E. D. T. Atkins, The crystal structure of gellan, *Carbohydr. Res.*, 175 (1988) 1–15.
35. R. P. Millane, R. Chandrasekaran, S. Arnott, and I. C. M. Dea, The molecular structure of kappa-carrageenan and comparison with iota-carrageenan, *Carbohydr. Res.*, 182 (1988) 1–17.
36. R. Chandrasekaran, R. P. Millane, and S. Arnott, Molecular structures of gellan and other industrially important gel-forming polysaccharides, in G. O. Phillips, D. J. Wedlock, and P. A. Williams, (Eds.), *Gums and Stabilisers for the Food Industry 4*, IRL Press, Oxford, UK, 1988, pp. 183–191.
37. R. Chandrasekaran, L. C. Puigjaner, K. L. Joyce, and S. Arnott, Cation interactions in gellan: An X-ray study of the potassium salt, *Carbohydr. Res.*, 181 (1988) 23–40.
38. R. P. Millane, T. V. Narasaiah, and S. Arnott, Molecular structures of xanthan and genetically engineered xanthan variants with truncated side chains, in V. Crescenzi, I. C. M. Dea, and S. Paoletti, (Eds.), *Biomedical and Biotechnological Advances in Industrial Polysaccharides*, Gordon and Breach, New York, USA, 1989, pp. 469–478.
39. R. P. Millane, E. U. Nzewi, and S. Arnott, Molecular structures of carrageenans determined by X-ray fiber diffraction, in R. P. Millane, J. N. BeMiller, and R. Chandrasekaran, (Eds.), *Frontiers in Carbohydrate Research-1: Food Applications*, Elsevier, London, 1989, pp. 104–131.
40. R. P. Millane and S. Arnott, Ordered water in hydrated solid state polysaccharide systems, in H. Levine and L. Slade, (Eds.), *Water Relationships in Foods: Advances in the 1980s and Trends for the 1990s, Advances in Experimental Biology and Medicine*, Vol. 302, Plenum Press, New York, USA, 1991, pp. 785–803.
41. S. Arnott, W. Bian, R. Chandrasekaran, and B. R. Manis, Lessons for today and tomorrow from yesterday—The structure of alginic acid, *Fiber Diffract. Rev.*, 9 (2000) 44–51.

SEVEN DECADES OF "ADVANCES"

Derek Horton

Department of Chemistry, Ohio State University, Columbus, Ohio, USA
Department of Chemistry, American University, Washington, District of Columbia, USA

I. Introduction and Background   13
    1. The Launching of *Advances* and Its Policies   13
    2. Carbohydrate Nomenclature and Indexing   14
    3. Carbohydrate Reference Books   16
    4. Founding of the Journal *Carbohydrate Research*   17
II. Editors' Prefaces for Volumes 1–70 of *Advances*   18
III. Concluding Remarks   207
    References   208

I. Introduction and Background

1. The Launching of *Advances* and Its Policies

This annual book series had its start during World War II when Walter J. Johnson and Kurt Jacoby, immigrants from Germany fleeing Nazi persecution, founded in 1942 the publishing house Academic Press, with its headquarters in New York City. Johnson was the business manager of the company while his brother-in-law Jacoby, a scholarly academic who had great editorial acumen, was able to pass along a spirit of intellectual participation upon which fruitful relationships between publishers and scientists rested.[1] Jacoby was an extraordinarily energetic editor, projecting both intelligence and zeal. He established close personal contacts in establishing new publishing ventures with such leading scientists as Joshua Lederburg, Hans Neurath, Linus Pauling, J. T. Edsall, Melville L. Wolfrom, W. Ward Pigman, and Sir John Kendrew, setting a standard for scientific excellence coupled with an indifference to short-term financial considerations. He died in 1968 at the age of 75 and was succeeded by Dr. James Barsky, a knowledgeable scientist who sustained Jacoby's commitment to scientific excellence. He served as President of Academic Press between 1978 and 1987. The company had in the meantime been acquired in 1969 by Harcourt Brace Publishers, and in the year 2000 Academic Press became an imprint of Elsevier.

In consultation with Wolfrom and Pigman soon after the founding of Academic Press, a plan was set up for a series of books to appear annually, featuring invited contributions covering a range of topics in the carbohydrate field. With a group of advisors comprising William Lloyd Evans, Hermann O. L. Fischer (son of Emil Fischer), R. Max Goepp, Sir Norman Haworth, and Claude S. Hudson, the first volume of *Advances* was launched, with Pigman and Wolfrom as the editors. A group of twelve contributing authors provided articles on a diversity of carbohydrate topics, with a notable first one from Claude S. Hudson on the Fischer cyanohydrin synthesis and one from Thomas J. Schoch that clearly demonstrated that starch is in fact a mixture of two different polysaccharides, one termed by Schoch as the A-fraction (nowadays amylose) and another the B-fraction (now amylopectin).

Difficulties in communication during the Second World War delayed the solicitation of manuscripts and in the production of the book, and the first volume appeared in 1945. In their Preface to that volume, Pigman and Wolfrom spelled out the basic policies and objectives of *Advances in Carbohydrate Chemistry*, with its scope intended to cover the simple sugars, glycosides, and polysaccharides, along with biochemical, industrial, and analytical aspects. Contributions were expected to be critical and integrating treatments of the particular field, and not mere literature reviews. For mature areas, the expectation was a definitive treatment with great historical accuracy to serve as a permanent record. For fields undergoing rapid development, the objective was to solicit a number of contributions reflecting different viewpoints.

In all instances the editors insisted that contributions be written in concise, grammatically correct English, in a style readily comprehensible by the average chemist or biochemist, and not just the specialist. Simple direct sentences, avoidance of jargon, and definition of specialized terms were expected, especially for the benefit of those readers not having English as their mother language.

2. Carbohydrate Nomenclature and Indexing

At the outset of the series there was considerable confusion in the literature concerning the naming of sugars and their derivatives, as they were outside the scope of the 1892 "Geneva" system for naming organic compounds based on parent alkane names, functional groups, and substituent groups. A multiplicity of trivial names for simple sugars, oligosaccharides, and polysaccharides abounded in the literature, with names often coined arbitrarily based on the source of the carbohydrate. There were many redundant names for the same compound. Semisystematic names based on modifiers to the trivial name were used inconsistently, and there were major differences even between the names for the same compound in the German, French, and English languages. Up to the 1940s, nomenclature proposals had been made by individuals; some of these were adopted in the scientific community, but others were not.

The International Union of Chemistry developed and expanded the Geneva nomenclature for organic compounds, but made little progress with the nomenclature of carbohydrates. The International Union of Pure and Applied Chemistry (IUPAC) Commission on Nomenclature of Biological Chemistry put forward a classification scheme for carbohydrates, but the new terms then proposed have not survived. However, in 1939, the American Chemical Society (ACS) formed a committee to look into this matter, since rapid progress in the field had led to various misnomers arising from the lack of guidelines.[2] This committee set out the foundations of modern systematic nomenclature for carbohydrates and derivatives: the numbering of the sugar chain, the use of D and L and α and β, and the designation of stereochemistry by italicized prefixes, with multiple prefixes for longer chains. The final report, prepared by M. L. Wolfrom, was approved by the ACS Council and published[3] in 1948.

The early volumes of *Advances* benefited greatly from the fact that *Chemical Abstracts*, a division of the American Chemical Society charged with abstracting and indexing the world's chemical literature, had its editorial offices on the campus of Ohio State University. These were directly proximal to the Ohio State Chemistry Department where Melville L. Wolfrom (and later Horton) had their laboratories. Chemical literature from all over the world was collected in abstract form in the volumes of *CA* by a large team of volunteer abstractors, and the content in various subject areas was coordinated by Section Editors, while the in-house staff and professional indexers produced the published volumes and indexes. The use of volunteer abstractors was finally phased out in 1994.

Dr. Leonard T. Capell, the Nomenclature Director and Executive Consultant at *Chemical Abstracts*, had overall responsibility for application of acceptable nomenclature standards in the *CA* indexes. He provided a most valuable service in compiling the indexes of many of the volumes of *Advances*, and in being conveniently accessible to Wolfrom (and later to this writer) for consultation on nomenclature matters, up until his retirement in 1964 and the move of the Chemical Abstract Service operation to a large complex to the north of the Ohio State campus. Dr. Capell compiled the index to the first volume of *Advances*, and most of the subsequent indexes up to Volume 61 in 1983. His contribution brought the high standard set for the *Chemical Abstracts* indexes and reflects the best usage of the day in the nomenclature of carbohydrates.

Not all problems were solved with the 1948 nomenclature document,[3] and different usages were encountered on the two sides of the Atlantic. The work of a joint British–American committee was published[4] in 1953 as "Rules for Carbohydrate Nomenclature," and a revised version was published in 1963 with endorsement by the ACS and by the Chemical Society in Britain.[5] The publication of this report led the IUPAC Commission on Nomenclature of Organic Chemistry, jointly with the IUPAC–IUB Commission on Biochemical Nomenclature, to issue, with broader international input, the "Tentative Rules for Carbohydrate Nomenclature, Part I, 1969," published in 1971/72 in several journals and later referred to as 1-Carb.[6] However, this document still did not reconcile various alternatives, especially concerning the stereodesignators for deoxy sugars and dicarbonyl sugars.

A large international group with 52 members, under the auspices of IUPAC–IUBMB, subsequently developed a comprehensive document, *Nomenclature of Carbohydrates, Recommendations 1996 (2-Carb)*, published in 1996, which reconciles earlier differences and has remained the basis for current usage.[7] In addition to basic definitions for sugars, glycosides, polysaccharides, and glycoconjugates, it includes recommendations for naming unsaturated sugars, branched-chain sugars, conformations of cyclic sugars, and polysaccharides, together with recommendations for naming a wide range of structurally modified carbohydrate derivatives. Also included are tabulations of trivial names for sugars and their systematic equivalents, and recommendations for the representation of sugar chains employing three-letter abbreviations for the monosaccharides. That document was also published in Volume 52 of *Advances*, and a version published on the World Wide Web has been routinely kept up to date with minor corrections.[8]

A short document *Nomenclature of Glycolipids, Recommendations 1997*, addressing specific questions in naming glycolipids, was developed by an IUPAC–IUBMB panel.[9] This document was also published in Volume 55 of *Advances* and also as a Web version.[9]

3. Carbohydrate Reference Books

Parallel with the *Advances* series, Ward Pigman, together with R. Max Goepp, planned a comprehensive monograph of the entire area of carbohydrate chemistry, *Chemistry of Carbohydrates*, which was published in 1948 by Academic Press.[10]

Pigman prepared an expanded and updated version, *The Carbohydrates, Chemistry, Biochemistry, and Physiology*, that was published in 1957, again under the Academic Press imprint.[11] Finally, a four-volume revision, *The Carbohydrates, Chemistry and Biochemistry, Second Edition*,[12] under the editorship of Pigman and Horton, and published again by Academic Press, appeared in 1970. Although now outdated, these volumes still contain important material on carbohydrates of permanent reference value. Also under the Academic Press imprint are the valuable eight-volume *Methods in Carbohydrate Chemistry* series founded by R. L. Whistler and M. L. Wolfrom,[13] and Aspinall's three-volume series *The Polysaccharides*.[14]

This author has provided a brief and selective overview of carbohydrate chemistry and biology, from antiquity to the present, citing seminal work of major pioneers along with key review articles of later developments.[15] Large compendia detailing much information on carbohydrates include the 1999 volume in the series *Comprehensive Natural Product Chemistry*[16] and the 2005 edition of *Polysaccharides: Structural Diversity and Functional Versatility*.[17] The 2008 volume *Glycoscience: Chemistry and Chemical Biology*[18] coordinates wide-ranging results on chemical and enzymatic approaches to synthesis, and the four-volume set *Comprehensive Glycoscience*,[19] published in 2007, is focused especially on biological aspects.

The *Dictionary of Carbohydrates*[20] lists some 24,000 individual carbohydrates, giving names, structures, sources, data, and literature references for each, together with a complementary electronic database. Another valuable reference work, which includes a wealth of information on carbohydrate derivatives important in biochemistry and biology, is the *Oxford Dictionary of Biochemistry and Molecular Biology*.[21]

Among the smaller single-author books on carbohydrates suitable for use as course texts may be noted Guthrie and Honeyman's *Introduction to Carbohydrate Chemistry*,[22] the Collins–Ferrier book *Monosaccharide Chemistry*,[23] and Stick's *Carbohydrates: The Essential Molecules of Life*,[24] which includes a brief treatment of biological aspects. Lehmann's book *Carbohydrates: Structure and Biology*[25] has the biological and biochemical aspects of carbohydrates as its central focus and is particularly useful for gaining a broad overview of the subject.

4. Founding of the Journal *Carbohydrate Research*

Twenty years after the launching of *Advances in Carbohydrate Chemistry*, Dr. Marc Atkins of the Elsevier Publishing Company approached Professor Allan B. Foster, of the University of London, with a view to setting up a new research journal devoted to all aspects of carbohydrates. Foster, my doctoral preceptor at the

University of Birmingham, invited me (by then a junior faculty member at Ohio State University) to join with him in this endeavor, and together with R. Stuart Tipson of the US National Bureau of Standards, Roger Jeanloz of Harvard University, and John Webber of Birmingham University, the new journal *Carbohydrate Research* came into being under the joint editorship of these five founding editors. That journal has burgeoned in size and scope during subsequent years and continues to flourish, with periodic changes in the editor lineup[26] up to the present day.

Taken together, the research journal *Carbohydrate Research*, the annual *Advances* series, and such monographs as the Pigman–Horton volumes and other books on carbohydrates, notably the *Methods in Carbohydrate Chemistry* series[12] founded by Roy L. Whistler and Melville L. Wolfrom, constitute a rich store of knowledge in the carbohydrate field, both for new advances and as a reference source for work in mature fields. A particularly important feature of this resource comes from the initiative of the Academic Press publishers, and later of Elsevier, to make full-text electronic access possible for the entire *Advances* series, and also for all articles in *Carbohydrate Research*.

The compilation that follows is based on a chronological collection of the Editor Prefaces of Volumes 1–69 of *Advances*. These prefaces record the evolution of the carbohydrate field over a period of seventy years, in its multifaceted manifestations of discovery and application. They also recount the knowledge and views of its major practitioners in their roles as contributing authors and provide a retrospect on their life stories as these leaders pass into history. For the benefit of today's and future readers, portraits of those carbohydrate scientists who built today's understanding of the field are included.

The first few Editor Prefaces provide no detail on the subject material of the respective volumes, and for these, there is included such information. From Volume 9 and later, the Prefaces feature the individual topics and their authors. It is hoped that this compilation will facilitate browsing of this rich archive of knowledge on carbohydrates in its many manifestations.

II. Editors' Prefaces for Volumes 1–70 of *Advances*

1. Volume 1

Editors' Preface.—The increasing tempo of research and the consequent increased specialization of research workers make it desirable to provide frequent reviews of important developments in carbohydrate chemistry, not only for carbohydrate chemists, but also for research workers in other fields, and industrial chemists and teachers. With this book there is begun the publication of a series of annual volumes entitled *"Advances in Carbohydrate Chemistry."* For each volume, invitations will be extended to selected research workers to prepare critical reviews of special topics in the broad field of the carbohydrates, including the sugars, polysaccharides, and glycosides. It is also the intention to cover, as far as the available space will permit, biochemical, industrial, and analytical developments. It is our plan to have the individual contributors furnish *critical*, integrating reviews rather than mere literature surveys and to have the articles presented in such a form as to be intelligible to the average chemist rather than only to the specialist.

Although the usual rules of the assignment of proper credit for developments will be followed, we do not believe it necessary to quote all past work done in a particular field and the contributions of a particular laboratory or group may be emphasized.

It may be found desirable to present several reviews of controversial subjects, particularly of those in fields undergoing a rapid state of development. In this way, different points of view will find expression. In addition to the presentation of topics covering recent advances, we are providing occasional articles which will review thoroughly special fundamental topics in carbohydrate chemistry. These articles will cover fields which have matured and will be quite complete from the historical standpoint. After a number of years, it is hoped that the aggregate of these articles will provide a fairly complete summary of carbohydrate researches.

The general policies of the *"Advances"* have been formulated by an Executive Committee consisting of W. L. Evans, H. O. L. Fischer, R. Max Goepp, Jr., W. N. Haworth, C. S. Hudson, and the two editors. It is a pleasure to announce that, beginning with the second volume, Dr. Stanley Peat of Birmingham University, England, will act as Associate Editor to solicit and edit contributions from the British Isles. It seems probable that an enlargement of the organization may be expected in the future.

Because the present volume is the first to be presented, we trust that the readers will not be too critical and will remember that the attainment of uniformity and the establishment of permanent policies will require some time and much consideration. The present international conflict has made the solicitation of manuscripts difficult and has provided many other difficulties. The cooperation shown by the contributors to the first volume is greatly appreciated.

We hope that the *"Advances"* will receive the wholehearted support of carbohydrate chemists in particular and of the chemical profession as a whole. Such support is necessary for the successful continuation of our work. We would be very glad to receive suggestions from the readers, of better ways in which we can serve the needs of carbohydrate chemists and of fields in need of review.

The support and encouragement given by the publishers in this undertaking are gratefully acknowledged. The index has been compiled by Dr. L. T. Capell. Mr. J. V. Karabinos has rendered valuable editorial assistance.

	THE EDITORS
Chicago, Illinois	W. W. Pigman
Columbus, Ohio	M. L. Wolfrom

The contributions to Volume 1 include an opening article from C. S. Hudson on the Fischer cyanohydrin synthesis, and others from Richtmyer (altrose), carbohydrate orthoesters (Pacsu), thio and seleno sugars (Raymond), carbohydrates of the cardiac glycosides (Elderfield), alditol metabolism (Carr and Krantz), nucleic acids (Tipson), the fractionation of starch (Schoch), starch esters (Whistler), cellulose esters (Fordyce), and plant "polyuronides" (Anderson and Sands).

2. Volume 2

Editors' Preface.—The first volume of this series was published during the war and contained chapters written only by American authors. In this second volume, we are especially pleased to be able to present several articles from the English school of carbohydrate chemists and one from France, thus making the *"Advances"* international in scope. Dr. Stanley Peat of Birmingham University, England, has been of great help in making arrangements for these and future articles. As communications between countries become facilitated, we hope to increase the number of countries that the participating authors represent.

We wish again to extend a cordial invitation to carbohydrate chemists to suggest topics in need of review and to suggest any way in which our contributions to the field of carbohydrates may be improved.

Dr. L. T. Capell has again compiled the subject index. The editorial assistance rendered by Edgar E. Dickey and Mary Grace Blair has been greatly appreciated. Dr. Claude S. Hudson has given invaluable aid in the editing of this and the preceding volume.

The journal abbreviations used are those employed by Chemical Abstracts. Unless otherwise noted, all temperature values are expressed in centigrade units.

	THE EDITORS
Appleton, Wisconsin	W. W. Pigman
Columbus, Ohio	M. L. Wolfrom

Volume 2 opens with an article by C. S. Hudson on melezitose and turanose, followed by a report from Peat on anhydro sugars. Additional reports include ascorbic acid analogues (F. Smith), hexitol synthesis (Lespieau), carbohydrate and fat metabolism (Deuel and Morehouse), "mucopolysaccharides" (Stacey), bacterial polysaccharides (T. H. Evans and H. Hibbert), pectic materials (Hirst and Jones), difructose anhydrides (E. J. McDonald), and cellulose ethers (Haskins).

3. Volume 3

Editors' Preface.—In presenting this third volume in the annual series of *Advances in Carbohydrate Chemistry*, the editors are pleased that the international scope of their endeavor has been further increased, the present volume containing contributions from England, Scotland, Sweden, and Germany in addition to those from the United States. It is a matter of satisfaction to note that Volume 1 has gone into its second printing. The editors have endeavored to maintain this publication on a high scholarly level and they thank the contributors for their full cooperation in this effort.

Our executive committee has lost the services of Dr. R. Max Goepp, Jr., by his tragic and untimely death in an airplane accident. His enthusiastic support will be greatly missed. We are proud to welcome as new members of the executive committee: Professor E. L. Hirst, of the University of Edinburgh; Dr. R. C. Hockett, scientific director of the Sugar Research Foundation, New York; and Professor C. B. Purves, of McGill University.

Dr. Stanley Peat, of the University of Birmingham, has continued to render assistance as associate editor for the British Isles. Dr. C. S. Hudson has again very materially assisted in the editing of this volume. The subject index has been compiled by Dr. L. T. Capell. Valuable editorial assistance has been rendered by Mr. Donald O. Hoffman.

	THE EDITORS
Appleton, Wisconsin	W.W. Pigman
Columbus, Ohio	M. L. Wolfrom

Once more C. S. Hudson is the opening contributor to this volume, writing on the history of Fischer's stereo-formulas, followed by Percival discussing sugar hydrazones and osazones, Fletcher dealing with cyclitols, and Helferich reporting on trityl ethers. Other articles included constituents of cane molasses (Sattler), oxidation of sugars by halogen (Green), constitution of cellulose (Compton), isotopic tracers in carbohydrate metabolism (S. Gurin), enzymatic degradation of starch and glycogen (Myrbäck), mycobacterial polysaccharides (Stacey and Kent), and an article by Lemieux and Wolfrom on the chemistry of streptomycin.

This last article was of great interest to a young chemistry student at Birmingham University named Derek Horton (this writer); it stimulated a life-long interest in amino sugars and later led to his moving to Columbus, Ohio to work with Wolfrom and subsequently to build a research career at Ohio State University.

4. Volume 4

Contributing Authors.—This volume, appearing in 1949 and edited by Ward Pigman and Melville Wolfrom, did not have an Editor's Preface, but gave the names and addresses of the contributors. The titles of their contributions are included here.

J. Böeseken, *The University, Delft, Holland*. The Use of Boric Acid for the Determination of the Configuration of Carbohydrates.

H. G. Bray and M. Stacey, *University of Birmingham, England*. Blood Group Polysaccharides.

Venancio Deulofeu, *Facultad de Ciencias Exactas, Fisicas y Naturales, Buenos Aires.. Argentina*. The Acylated Nitriles of Aldonic Acids and Their Degradation.

R. Lohmar , *Northern Regional Research Laboratory, U. S. Department of Agriculture, Peoria, Illinois* and R. M. Goepp, Jr. (deceased October 1946), *Atlas Powder Company, Wilmington, Delaware*. The Hexitols and Some of Their Derivatives.

E. E. Harris, *Forest Products Laboratory, U. S. Department of Agriculture, Madison, Wisconsin*. Wood Saccharification.

C. S. Hudson, *National Institutes of Health, Bethesda, Maryland*. Apiose and the Glycosides of the Parsley Plant.

J. K. N. Jones, *University of Manchester, England*, and F. Smith, *University of Birmingham, England*. Plant Gums and Mucilages.

I. Levi, *Charles E. Frosst and Company, Montreal, Canada*, and C. B. Purves, *McGill University, Montreal, Canada*. The Structure and Configuration of Sucrose (α-D-Glucopyranosyl-β-D-fructofuranoside).

Carl Neuberg, *Department of Chemistry, Polytechnic Institute of Brooklyn, New York*. Biochemical Reductions at the Expense of Sugars.

L. F. Wiggins, *University of Birmingham, England*. The Utilization of Sucrose.

5. Volume 5

Editors' Preface.—The editors of Volume V wish to express their appreciation of the excellent service that has been rendered by Professor W. W. Pigman and Professor M. L. Wolfrom in the editing of the earlier volumes of this series. They have generously consented to assist in the publication of later volumes in the capacity of members of a "Board of Advisors," which now takes the place of the "Executive Committee" of the earlier volumes.

During the past year, we have suffered a great loss through the death of Sir Walter Norman Haworth. His assistance was invaluable in the organization of the group that initiated the publication of the *Advances* and he gave much of his time to the details of the solicitation of articles during the remaining years of his life. We pay homage to the memory of this great man and genial friend.

The editorial policy of the publication will continue in its past form. To quote from Volume I: "It is our plan to have the individual contributors furnish critical, integrating reviews rather than mere literature surveys, and to have the articles presented in such a form as to be intelligible to the average chemist rather than only to the specialist." Invitations will be extended to selected research workers to prepare critical reviews of special topics in the broad field of the carbohydrates, including the sugars, polysaccharides, and glycosides. It is also the intention to include biochemical and analytical developments in the carbohydrate field as well as critical reviews of important industrial advances.

Criticisms and suggestions of any kind are respectfully solicited.

THE EDITORS
Bethesda, Maryland   C. S. Hudson
Philadelphia, Pennsylvania   S. M. Cantor

Subjects in Volume 5 include desulfurization by Raney nickel (Fletcher and Richtmyer), enzymatic synthesis of sucrose (Hassid and Doudoroff), principles underlying enzyme specificity (Gottschalk), enzymes acting on pectic substances (Kertez and McColloch), crystallinity of cellulose (Nickerson), commercial production of dextrose (Dean and Gottfried), the methyl ethers of D-glucose (Bourne and Peat), anhydrides of the pentitols and hexitols (Wiggins), action of alpha amylases (Caldwell and Adams), and xylan (Whistler).

6. Volume 6

Editors' Preface.—We regret to report the death of our esteemed collaborator, Dr. Edmund George Vincent Percival, Reader in Chemistry, the University of Edinburgh, Scotland, on September 27th, 1951, at the age of forty-four. The importance of his contributions to the progress of carbohydrate research is universally recognized; his ability in teaching and his friendliness endeared him to a wide circle of students and colleagues, who mourn his passing. His aid to this publication, both as a contributor and as a member of its Board of Advisors, is here recorded with deepest appreciation.

In addition to the usual author and subject indexes for volume 6, there is included also a cumulative subject index for the preceding five volumes. This cumulative index is offered particularly to research workers as an aid in tracing matters back to the original publications in specialized journals.

We are pleased to announce that Dr. M. L. Wolfrom will rejoin the editorial staff, beginning with volume 7.

	THE EDITORS
Bethesda, Maryland	C. S. Hudson
Philadelphia, Pennsylvania	S. M. Cantor

Subjects in Volume 6 include the methyl ethers of D-galactose (D. J. Bell), oligosaccharide synthesis (Evans, Reynolds, and Talley), formation of furan compounds from hexoses (Newth), cuprammonium glycoside complexes (Reeves), the chemistry of ribose (Jeanloz and Fletcher), the 2-aldopolyhydroxyalkyl) benzimidazoles (Richtmyer), granular adsorbents for sugar refining (Barrett), aconitic acid as a byproduct of sugar manufacture (Miller and Cantor), Friedel–Crafts and Grignard processes in carbohydrates (Bonner), and the nitromethane and 2-nitroethanol syntheses (Sowden).

Appearing for the first time in *Advances* is an obituary article, written by E. L. Hirst, which commemorates the life and career of Sir Norman Haworth. In later volumes in the series, such articles have continued to feature the lives of prominent researchers in the carbohydrate field, especially those who had made notable contributions to *Advances*.

Sir (Walter) Norman Haworth

7. Volume 7

Volume 7 of *Advances in Carbohydrate Chemistry* did not have an Editors' Preface. The joint editors of the volume were Melville L. Wolfrom of the Department of Chemistry, Ohio State University, Columbus, Ohio; Claude S. Hudson of the National Institutes of Health, Bethesda, Maryland; and Sidney M. Cantor of the American Sugar Refining Company, Philadelphia, Pennsylvania.

Listed as Associate Editors for the British Isles were Stanley Peat of University College of North Wales, Bangor, Caernarvonshire, Wales, and Maurice Stacey of The University, Birmingham, England.

Subjects (and authors) for Volume 7 were the methyl ethers of the aldopentoses and of rhamnose and fucose (Laidlaw and E.G.V. Percival), the 1,6-anhydrohexofuranoses (Dimler), fructose and its derivatives (Barry and Honeyman), psicose, sorbose, and tagatose (Karabinos), acetals and ketals of the tetritols, pentitols, and hexitols (Barker and Bourne), the glycals (Helferich), the 2-amino-2-deoxy sugars (Foster and Stacey), and the size and shape of polysaccharide molecules (Greenwood).

8. Volume 8

Editors' Preface.—The sudden death of Claude S. Hudson on December 27, 1952, in his home at Washington, D.C., removes from carbohydrate chemistry one of its great and inspiring leaders. Thus, in the space of a few years, have passed away three great pioneers in this field, W. N. Haworth (1950), J. C. Irvine (1952), and C. S. Hudson (1952). The last had been a guiding spirit for "*Advances in Carbohydrate Chemistry*" since its inception in 1944. For the past four years he had been an active editor and, indeed, since his retirement in January, 1951, the editorship of the "*Advances*" had occupied the greater portion of his time. Dr. Hudson set up exacting standards for his own writing and research and held to a high quality of scholarship in these endeavors. He laid down the policy that the attempt should be made to hold all of the chapters in "*Advances in Carbohydrate Chemistry*" to the criteria established in his own writing, while maintaining the integrity of the authors concerned and changing nothing without their full consent. The enforcement of such a policy is not without attendant difficulties; its degree of success may be judged by the readers of these volumes. The manuscripts for the present edition were in the hands of Dr. Hudson at the time of his demise and all had received his editorial attention.

Carbohydrate nomenclature has been an ever-present problem in this series. It has recently been the subject of rather extensive consultations between representatives of the American and British carbohydrate chemists, and the final results have appeared in printed form, *Chem. Eng. News,* 31, 1776 (1953) and *J. Chem. Soc.,* 5108 (1952). Meanwhile, the present volume represents rather a transition stage in this development, particularly as regards the use of the *O*-substitution indication, which has been employed only in part. More-uniform usage may be expected in the future.

Dr. R. Stuart Tipson has rendered important service in the editing of this volume and has prepared the index.

Columbus, Ohio   M. L. Wolfrom

Contributions to Volume 8 included Sugihara writing on relative reactivity of hydroxyl groups in carbohydrates, the "2-desoxy sugars" (Overend and Stacey), sulfonic esters of carbohydrates (Tipson), methyl ethers of D-mannose (Aspinall), synthesis of D-glucuronic acid (Mehltretter), D-glucuronic acid in metabolism (Bray), the substituted-sugar structure of melezitose (Hehre), the composition of cane juice and cane final molasses (Binkley and Wolfrom), and seaweed polysaccharides (Mori).

9. Volume 9

Editors' Preface.—In this Volume, we are pleased to present contributions from a number of the younger, carbohydrate-trained chemists. With the growth of the rapidly enlarging area of chemical publication, it will be necessary to depend more upon the efforts of these selected young men who can still find the time to tabulate and correlate the mountainous literature which confronts the modern investigator.

Dr. G. O. Aspinall continues our series on the methyl ethers of the sugars. Modern interpretations of reaction mechanism are applied to carbohydrate derivatives by Dr. C. E. Ballou and by Dr. R. U. Lemieux, and those considerations are developed which led to the recent synthesis of sucrose by Lemieux and Huber.

Dr. G. N. Kowkabany offers a summary of the rapidly developing field of paper chromatography, which is revolutionizing the analysis of sugar mixtures. Professor Dexter French presents another of our chapters on the nature of a group of plant sugars, in this case the raffinose family of oligosaccharides. Dr. M. Grace Blair discusses the hydroxyglycals.

From time to time, we have offered chapters relating to the physiology of the carbohydrates, and Professor R. S. Teague herein develops such a topic for the animal conjugates of D-glucuronic acid. This contribution extends that of Bray in the preceding Volume, and so emphasizes a subject of unusual current interest.

Industrial aspects of the carbohydrate field are considered by R. W. Liggett and V. R. Deitz, and by J. V. Karabinos and Marjorie Hindert, all of whom have had experience in the industries upon which they report.

The attempt has been made to bring all of the carbohydrate nomenclature employed in this Volume into conformity with the rules published in *Chemical and Engineering News,* 31, 1776 (1953).

The long and highly valued association of the late Dr. Claude S. Hudson with our publication terminates herewith in a short obituary notice.

Dr. R. Stuart Tipson has again rendered inestimable service in the editing and indexing.

Columbus, Ohio   M. L. Wolfrom

Claude Silbert Hudson

10. Volume 10

Editor's Preface.—With this Volume, "*Advances in Carbohydrate Chemistry*" comes of age through the completion of a series of ten issues. Herein, modern conformational analysis is applied to the carbohydrate field by J. A. Mills (Adelaide). The ever-recurrent and always puzzling subject of nitrogen chemistry is elaborated in a discussion of the glycosylamines and their rearrangement products by G. P. Ellis with J. Honeyman (London) and by J. E. Hodge (Peoria).

The preparation and reactivity of the useful glycosyl halides is presented by L. J. Haynes (Edinburgh) and F. H. Newth (Cambridge). W. W. Binkley (Columbus) summarizes the present status of column chromatographic technique as applied to the sugar group. Our series of chapters on the methyl ethers is augmented by G. G. Maher (Clinton, Iowa).

Polysaccharide chemistry is represented by a chapter on the non-cellulosic components of wood from the pen of W. J. Polglase (Vancouver) and by one on the biochemically significant subject of heparin from A. B. Foster and A. J. Huggard (Birmingham and Columbus). These, together with an obituary of the late, esteemed Dr. E. G. V. Percival of Edinburgh, complete this Volume and are offered as a contribution to the summarizing of progress in the ever-growing subject of carbohydrate chemistry.

Columbus, Ohio   M. L. Wolfrom

E. G. V. Percival

Edward George Vincent Percival

11. Volume 11

Editors' Preface.—In this Volume, we present a discussion by J. M. Bobbitt on the application to carbohydrates of the powerful tool of periodate cleavage. The subject of the osones is reviewed by S. Bayne and J. A. Fewster, and that of the β-ketonic ester–monosaccharide condensation products by F. García González. The chemistry of kojic acid is developed by A. Beélik.

Modern biochemical aspects of the carbohydrates are delineated in a discussion by L. Hough and J. K. N. Jones on the biosynthesis of the monosaccharides. The branched-chain sugars are becoming more significant, especially in connection with the newer antibiotic substances, and their chemistry and occurrence are summarized by F. Shafizadeh. This author also proffers a solution to their puzzling nomenclatural requirements, framed within the compass of the established rules. The problem presented by a bifunctional substituent attached both to a carbon atom of the main chain and to one in a side-chain (formula XII, page 266) has not hitherto been encountered in organic chemistry; the reader can determine whether the solution offered is a sufficient one.

Research activity is tremendous on the highly significant group of compounds known as the nucleic acids. Consequently, although this topic was covered in Volume I by Tipson, it has been deemed appropriate to bring the subject to a current status, and G. R. Barker was selected to perform this assignment. C. T. Greenwood offers a contribution to the physico-chemistry of natural high polymers in his discussion on Aspects of the Physical Chemistry of Starch.

An obituary of the late Kurt H. Meyer was written by one of his students, R. W. Jeanloz.

Attention is drawn to a special feature of this Volume—inclusion of a *generalized* Cumulative Index to Volumes 1–10. It is believed that use of this Cumulative Index, in conjunction with the detailed Cumulative Subject Index to Volumes 1–5 (in Volume 6) and the detailed, individual Subject Indexes to Volumes 6–10, will afford the reader speedy access to sources of detailed information in the various Volumes. This solution to the indexing problem was chosen for economic reasons.

A Cumulative Author Index to Volumes 1–10 replaces the previous listing of Contents of Volumes.

Columbus, Ohio	M. L. Wolfrom
Pittsburgh, Pennsylvania	R. Stuart Tipson

Kurt H. Meyer

12. Volume 12

Editors' Preface.—With this Volume, the *Advances in Carbohydrate Chemistry* enters upon its twelfth year of publication with chapters shared about equally between British and American writers. Foster (Birmingham) continues our series on modern carbohydrate-separation techniques with a contribution on zone electrophoresis. After lying dormant for many years, the theory and practice of saccharinic acid formation is undergoing a current revival, reported on by Sowden (Washington University).

Modern instrumentation has made the infrared absorption region of molecules readily available to the chemist. The complexities found in this spectral area with carbohydrate substances are still largely uninterpretable, but a start has been made which has been summarized by Neely (Dow Chemical Co.).

Topics mainly biochemical in nature are treated by French (Iowa State), Manners (Edinburgh), and by Whistler and Olson (Purdue); these are, respectively, the Schardinger dextrins, the glycogens, and hyaluronic acid. The fundamental organic chemistry of sugars is represented by chapters on sugar nitrates by Honeyman and Morgan (London), benzyl ethers by McCloskey (Pasadena), and simple glycosides by Conchie, Levvy, and Marsh (Rowett Research Institute, Scotland).

One of the editors (R. S. T.) has contributed a sketch of the life and work of the pioneer biochemist Phoebus A. Levene.

Columbus, Ohio	M. L. Wolfrom
Pittsburgh, Pennsylvania	R. Stuart Tipson

Phoebus Aaron Theodore Levene

13. Volume 13

Editors' Preface.—Volume 6 of this series contained a review of the pioneer work on sugar-ring conformational analysis carried out by Richard E. Reeves. This general topic has received considerable attention in later volumes and is herein extended by Shafizadeh to the formation and cleavage of the sugar oxygen rings.

Current emphasis on sugar–alkali reactivity is reflected in three chapters—by Speck, by Crum, and by Whistler and BeMiller. Crum has summarized the life-work of the late J. W. E. Glattfeld (and associates) on the four-carbon saccharinic acids. These outstanding researches of Glattfeld, a Nef student, have not hitherto received the recognition they merited.

Application of the long-known formazan reaction to sugar hydrazone structures is detailed by Mester, whose review, incidentally, serves to emphasize that the acyclic forms of sugar derivatives exist and cannot be ignored. Our series of chapters on the methyl ethers of the monosaccharides is continued with a contribution by Jeanloz on such derivatives of the amino sugars. The current interest in topics related to nucleic acid chemistry is reflected in a discussion of glycosyl ureides by Goodman.

The chemical nature of the sialic (nonulosaminic) acids is at last becoming clarified and this subject is summarized by Zilliken and Whitehouse. Stoloff contributes a chapter on some aspects of the polysaccharide hydrophilic colloids, and Caesar has written a fascinating account of the development of starch nitrate as an explosive.

Finally, the pioneer biochemist, Carl Neuberg—who, indeed, coined the term "biochemistry"—is memorialized in an obituary by one of his students, Professor F. F. Nord.

The editors finally admit defeat in their struggles to delineate sex by elaborating the first names of female scientists. As one of our esteemed reviewers has noted, this effort has been laudable but has led to a number of interesting sex reversals; thus, what American editor could be sure of the sex indicated by many of the Scandinavian or Hungarian given names, or would know that "Evelyn" is an established name for a British male? Accordingly, only initials are employed in this volume.

Columbus, Ohio   M. L. Wolfrom
Washington, D.C.   R. Stuart Tipson

Carl Neuberg

14. Volume 14

Editors' Preface.—In Volume 14 of this series, we bring up to date three significant topics previously covered in other volumes. These concern areas of research in which there has been exceptional activity with resultant progress; they are, respectively, the amino sugars (Foster and Horton), the hemicelluloses (Aspinall), and the inositols (Angyal and Anderson).

The topic of nucleic acids has twice been summarized in this series, but the unusually fertile studies current in this sector of carbohydrate investigation call for renewed coverage, and the chapter on pyrimidine nucleosides (Fox and Wempen) is part of a projected series of chapters covering subdivisions of this rapidly expanding subject.

The action on sugars of the oxidizing agent lead tetraacetate is delineated by Perlin, as a continuation of our series of chapters devoted to the action of oxidizing agents structurally specific toward carbohydrates. From time to time, we have offered summaries relating to enzymes affecting carbohydrates, and Levvy and Marsh herein describe β-glucuronidase.

Ellis gives a succinct summary of the present moot status of the Maillard (non-enzymic) browning reaction effected between sugars and amino acids. Finally, an obituary of the late Geza Zemplén is provided by Mester, one of the later coworkers of this distinguished disciple of Emil Fischer.

For the past six years, the Subject Indexes have been compiled by one of us (R. S. T.); the task has this year been assumed by Dr. Joseph D. Moyer. Dr. D. Horton assisted in the final editing.

Columbus, Ohio M. L. Wolfrom
Washington, D.C. R. Stuart Tipson

Geza Zemplén

The present author (Horton) notes that this volume marks his first contribution of an article to *Advances,* as well as his first introduction to editorial work in the series. Wolfrom was visiting England in 1956, and invited Foster and his graduate student Horton (who was working on amino sugars) to contribute the article that appeared in this volume.

15. Volume 15

Editors' Preface.—This volume completes fifteen issues in the series initiated in 1945. A resume of the life and work of the late Emil Heuser is herein provided by L. E. Wise, one of his associates, in the U. S., in a career embracing both sides of the Atlantic. Capon and Overend comment on the general topic of the physicochemical properties of sugars, while Bouveng and Lindberg collate methods currently employed in the assignment of structure to polysaccharides.

Specific instances of polysaccharide structure and properties are taken up by Neely for the bacterial polysaccharides of the dextran group, and by Foster and Webber for chitin, the widely occurring, encrusting polysaccharide. An extensive tabulation of the amino sugars and their derivatives is made by Horton, as a modernized replacement for the one in Volume 7 and as an appendix to the chapter on Aspects of the Chemistry of the Amino Sugars by Foster and Horton in the preceding volume of this series.

An article by Hough, Priddle, and Theobald delineates the chemistry and rather complex nomenclature of the sugar carbonate and thiocarbonate esters and indicates that these esters might well be employed more extensively as "blocking groups" in synthetic efforts.

The subject of the carbohydrate components of bacteria is always fascinating; it is reviewed again, by Davies and by Jonsen and Laland, in two closely related chapters. Finally, Sprinson writes authoritatively on the biosynthesis of aromatic amino acids. It is shown that the benzene ring originates in carbohydrate material, thus placing the aromatic compounds as a subsection under the carbohydrates (from the bioorganic viewpoint).

The Subject Index has been compiled by Dr. Robert Barker.

The editors record with sorrow the death on March 9, 1960 of Professor Hermann O. L. Fischer, a member of our Board of Advisors since the inception of this series.

Columbus, Ohio M. L. Wolfrom
Washington, D.C. R. Stuart Tipson

Emil Heuser

16. Volume 16

Editors' Preface.—This sixteenth Volume of the *Advances in Carbohydrate Chemistry* continues with its task of providing comprehensive reviews on matters of interest in the general chemistry of the carbohydrates. A significant problem is treated by Phillips (Cardiff)—the effects produced in carbohydrates by ionizing radiation—a topic which is in its infancy and which can be expected to undergo extensive development.

Fluorine chemistry is likewise a modern subject undergoing intensive study, and some phases of its application to carbohydrates are detailed by Bonner (London) and by Micheel and Klemer (Münster). The effects of various glycol-splitting reagents on the carbohydrates have been summarized in previous issues of this Series, and recent results now allow an elaboration of the ring structures of the dialdehydes produced from the pyranoid sugar rings by periodate ion (Guthrie, Manchester).

In the early issues of the *Advances*, the late Claude S. Hudson initiated a set of articles on single sugars (or simple groups of sugars), and lactose was one of those selected for discussion. This account, started in 1954 by Whistler (Purdue) but never completed to his satisfaction, has at last been finished by Hough and Clamp (Bristol).

Biochemical aspects have been treated authoritatively by Lederer (Paris), who reports on the interesting new sugars found in the glycolipids of the acid-fast bacteria; Wallenfels and Prakash Malhotra (Freiburg i. B.) detail the fascinating subject of the first isolation and crystallization of a simple glycosidase; and Deuel and associates (Zürich) discuss the carbohydrate residues isolable from the soil.

In the first Volume of this Series, T. J. Schoch described a fractionation of starch by which he firmly established the existence of the amylose and amylopectin fractions. His process remained a laboratory procedure only, but, recently, Dutch chemists have developed a large-scale fractionation of potato starch, and pure amylose is now obtainable in commercial quantities; the new process is herein described by Muetgeert (Delft).

Finally, an obituary of the late Harold Hibbert is offered by one of his former associates. The Subject Index has been prepared by Dr. R. David Nelson.

Columbus, Ohio	M. L. Wolfrom
Washington, D.C.	R. Stuart Tipson

Harold Hibbert

17. Volume 17

Editors' Preface.—We herewith offer Volume 17 of the *Advances in Carbohydrate Chemistry*. In the first volume of this series, the late Claude S. Hudson presented a review of the higher-carbon sugars which has now been brought up to date by J. M. Webber of Birmingham, England. J. A. Montgomery and H. Jeanette Thomas, of Birmingham, Alabama, continue with our series on nucleosides and nucleotides, the first of which was written by R. S. Tipson for Volume 1.

It is a pleasure to place on record the authoritative summary of many years of work, by T. Reichstein and coworkers, on the strange sugars found in the cardiac glycosides and first noted by Heinrich Kiliani. R. W. Bailey and J. B. Pridham cooperate between New Zealand and London to provide a much-needed review of the many oligosaccharides disclosed by the newer isolative methods.

O. Theander of Stockholm furnishes a wide-ranging review of the dicarbonyl sugars and their derivatives, while K. Heyns and H. Paulsen of Hamburg concentrate on one type of oxidation procedure, namely, that employing oxygen with a platinum catalyst.

D. J. Manners of Edinburgh ably summarizes the present status of the enzymic synthesis and degradation of starch and glycogen, in an article wherein the Editors regret only the babel of initials employed for the designation of many of the starch hydrolases, in contrast to the more-meaningful names utilized for the glycogen hydrolases; hopefully, the chapter may awaken the workers concerned to this deplorable situation.

Finally, J. C. Sowden closes a brilliant classical period in sugar chemistry by a review of the life and work of the late, beloved Hermann O. L. Fischer. The Subject Index has been prepared by Dr. Edward J. Hedgley.

Columbus, Ohio	M. L. Wolfrom
Washington, D.C.	R. Stuart Tipson

Hermann O. L. Fischer

18. Volume 18

Editors' Preface.—As we near the twentieth volume in this Series, we are pleased to note that Volume 18 has an aspect even more international than usual. Hassan El Khadem of Alexandria, Egypt, outlines the chemistry of the osotriazoles, derivatives which contain an unusual heterocycle established in the sugar series by Hann and Hudson.

D. Horton and D. H. Hutson of Ohio State bring to currency the subject of the thio sugars, originally described by A. L. Raymond in our very first volume. The trehaloses are detailed by G. G. Birch of Weybridge, England, and L. J. Haynes of Jamaica ably discusses C-glycosyl compounds that occur naturally. Continuing our presentation of useful techniques, we provide a summary of carbohydrate electrophoresis by H. Weigel of London.

The photochemistry of carbohydrates has intrigued and baffled chemists over the years; it is herein elaborated in its modern phases by G. O. Phillips of Cardiff. The chemistry of the carbohydrates has long had significant biochemical and physicochemical facets, and the subject of the physical properties of solutions of polysaccharides is presented by W. Banks and C. T. Greenwood of Edinburgh.

Our currently developing insight into the biosynthesis of saccharides through nucleotide intermediates is treated authoritatively by Elizabeth F. Neufeld and W. Z. Hassid of Berkeley, and another contemporary biochemical field, that of the many rare amino sugars found in antibiotic substances, is discussed by J. D. Dutcher of New Jersey.

Continuing our brief memoirs on carbohydrate chemists, J. E. Courtois of Paris contributes a sympathetic summary of the life of the French chemist Émile Bourquelot.

The Subject Index has been prepared by R. David Nelson, who is responsible for the indexing of carbohydrates for *Chemical Abstracts*.

The editors record with sorrow the death on April 14, 1963, of Professor John C. Sowden, a member of our Board of Advisors since 1950.

Columbus, Ohio   M. L. Wolfrom
Washington, D.C., April 1963   R. Stuart Tipson

Émile Bourquelot

19. Volume 19

Editors' Preface.—The nineteenth volume in this Series presents four chapters on modern instrumental techniques as applied to carbohydrates. These are crystal-structure analysis by G. A. Jeffrey and R. D. Rosenstein (Pittsburgh), infrared spectroscopy by H. Spedding (Alberta), nuclear magnetic resonance by L. D. Hall (Ottawa, The University), and gas–liquid chromatography by C. T. Bishop (Ottawa, National Research Council). Such techniques are rapidly revolutionizing the investigative methods of structural organic chemistry and are of especial interest in the carbohydrate field.

Modern reaction mechanisms involving carbohydrate structures are discussed: by G. J. Moody (Cardiff), as they apply to the action of hydrogen peroxide, an old reagent in this area; by E. F. L. J. Anet (Sydney, Australia), in regard to the significant degradation of carbohydrates to dicarbonyl compounds; and by D. M. Jones (Manchester, Shirley Institute), for that most investigated entity in organic chemistry, the cellulose macromolecule.

Studies that are, essentially, purely structural are delineated by T. E. Timell (Syracuse), and by M. J. How, J. S. Brimacombe, and M. Stacey (Birmingham). The former chapter deals with the polysaccharides accompanying cellulose in wood, an area long held to be mysteriously unknown, but now being well revealed by modern techniques. Because of its length, this chapter has, perforce, been divided in two. Part II, which deals with the hemicelluloses in the wood of gymnosperms, will appear in Volume 20.

The final chapter, by Stacey and associates, blazes a path through that intricate maze of polysaccharide structures constituting the bulk of the various pneumococcal capsular materials. In these investigations, the powerful methods of immunochemistry, largely established by Michael Heidelberger, have been utilized to their utmost. The Editors trust that this collection of essays may prove of particular interest and value.

The Subject Index has again been compiled by R. David Nelson of the Chemical Abstracts Service. The obituary in this Volume, by Derek Horton, pays tribute to Alva Thompson, long associated with one of the Editors, and respected and beloved by all with whom he came in contact.

Columbus, Ohio
Washington, D.C. August, 1964

M. L. Wolfrom
R. Stuart Tipson

Alva Thompson

20. Volume 20

Editors' Preface.—The editors are proud to note that this volume marks the twentieth issue in this serial publication. It is, therefore, in the nature of a Jubilee Volume. One of the editors has been with the publication since its inception, except for a short period during which he served on the Board of Advisors. Although not officially an editor of the first four volumes, the late Claude S. Hudson served as the guiding spirit for these volumes from their inception, and this interest was maintained throughout the rest of his life. Hudson laid down the editorial principles which the editors have ever since endeavored to follow and which were strongly supported at the time by Hudson's good friend, the late Sir Norman Haworth.

In this volume, G. E. McCasland (San Francisco) discusses the deoxyinositols, and especially shows how the modern technique of nuclear magnetic resonance has aided in their structural elucidation. R. J. Ferrier (Birkbeck, London) reviews the present status of olefin chemistry as applied to the sugars, and thus delineates an area soon due for much further development. H. El Khadem (Alexandria, Egypt) ably proves that the chemistry of the sugar "osazones" can still provide much of current interest.

The subject of the sulfate half-esters of the simple sugars is brought up to date by J. R. Turvey (Bangor, Wales). A. N. de Belder (Holloway, London) provides a much-needed summary of the structure and reactivity of the important and useful cyclic acetals of the glycosides and aldoses. F. García Gonzalez and A. Gómez Sanchez (Seville, Spain) extend a previous chapter (in Volume 11) by discussing reactions of the amino sugars with β-dicarbonyl compounds.

Some aspects of the carbohydrate chemistry of plant phenolics are delineated by J. B. Pridham (Holloway, London), and L. J. Haynes (Kingston, Jamaica) offers a short addendum to his previous chapter (in Volume 18) on the interesting C-glycosyl compounds of plants. T. E. Timell (Syracuse, New York) concludes his discussion of the polysaccharides accompanying cellulose in wood with Part II of a chapter that started in Volume 19.

Finally, an obituary of John C. Sowden is furnished by his life-long friend S. M. Cantor.

The editors record with sorrow the death on February 1, 1965, of Professor Fred Smith, a contributor of two articles to this serial publication, and the death on September 30, 1965, of Professor C. B. Purves, a member of our Board of Advisors since 1948.

Columbus, Ohio
Washington, D.C., October, 1965

M. L. Wolfrom
R. Stuart Tipson

John C. Sowden

21. Volume 21

Editors' Preface.—The editors herewith present the twenty-first volume in this serial publication. To celebrate our "coming of age," we are proud to offer a review of the contributions of Emil Fischer to carbohydrate chemistry, by one of his students, Professor Karl Freudenberg. In translation, some of the fine expression and style of the original German may have been lost, yet the review is nevertheless an outstanding evaluation of Fischer's contributions to the fundamentals of modern carbohydrate chemistry.

Current organic chemistry is ever utilizing new instrumentations and techniques that have originated in physics and have been perfected by the modern instrument fabricator. The newest instrument to impinge upon the carbohydrate field is the mass spectrometer, and a review of current work on its use in this area is made by Kochetkov and Chizhov (Moscow).

This volume contains two chapters which update topics presented in earlier ones. The chemistry of the deoxy sugars has been expanded considerably since the review in Volume 8 by Overend and Stacey (Birmingham), as is attested by the chapter by Hanessian (Ann Arbor). The article on synthetic cardenolides, or cardiac glycosides, by Zorbach and Bhat (Georgetown) is an extension of related topics previously reviewed by Elderfield (Volume 1) and by Reichstein and Weiss (Volume 17).

The discussion of chemical synthesis of polysaccharides, by Goldstein and Hullar, is another contribution to those chapters on carbohydrate polymers which have appeared in this serial publication. Green (Appleton) reviews the generally neglected topic of glycofuranosides.

Inorganic chemistry is included in the chapter on complexes of alkali metals and alkaline-earth metals with carbohydrates by Rendleman (Peoria). Finally, two topics in biochemistry are reviewed by Archibald and Baddiley (Newcastle) and by Hilton (Honolulu), the former being concerned with the teichoic acids, and the latter with the effects of plant-growth substances on carbohydrate systems in plants.

The Subject Index for this as well as for the preceding volume has been prepared by Dr. L. T. Capell, long associated with *Chemical Abstracts* and an internationally recognized authority on organic nomenclature.

Columbus, Ohio	M. L. Wolfrom
Gaithersburg, Maryland November, 1966	R. Stuart Tipson

Emil Fischer

22. Volume 22

Editors' Preface.—We are pleased to offer in Volume 22 of this serial publication a chapter on fructose by Verstraeten (Heverlee, Belgium) which brings to date earlier chapters on this important sugar (Volume 7, 1952) and its dianhydrides (Volume 2, 1946). Halogenated carbohydrates constitute some of the earliest known carbohydrate derivatives, and this significant topic has been reviewed by Barnett (Southampton).

Modern concepts of the acid-catalyzed hydrolysis of glycosides arc presented by BeMiller (Southern Illinois), and a review of the established process of acetolysis is offered by Guthrie and McCarthy (Sussex). We are especially proud to have, in this volume, a chapter entitled "Neighboring-Group Participation in Sugars" by Goodman (California), one of the leading authorities in this important aspect of organic chemistry.

In our series of chapters bearing on the currently popular subject of nucleic acids and their components, we have in this volume a monumental one on mononucleotides by Ueda and Fox (Sloan–Kettering, New York).

Greenwood (Edinburgh) contributes a treatise on the thermal degradation of starch, and Marchessault and Sarko (Syracuse) have provided a general review of the X-ray structure of polysaccharides.

An obituary of the late Fred Smith, one of the truly important carbohydrate workers, is given by one of his close associates, Rex Montgomery. It may be noted that the authors herein are about evenly divided between European and United States sources.

The Subject Index was prepared by Dr. Leonard T. Capell.

Columbus, Ohio	M. L. Wolfrom
Kensington, Maryland October, 1967	R. Stuart Tipson

1911–1965

Fred Smith

23. Volume 23

Editors' Preface.—In the twenty-third volume of this serial publication, we offer a long-delayed but notable contribution, by Pigman and Isbell (New York and Washington), to the modem evaluation of the classical phenomenon of sugar mutarotation; these authors have published much significant work in this area. Ball and Parrish (Natick) update the chapter on carbohydrate sulfonates, written by Tipson, that appeared in Volume 8 (1953). Because of their length, each of these chapters has been divided in two; Part II of each will appear in a succeeding volume.

Rosenthal (Vancouver) summarizes his many publications on the application of the oxo reaction to the carbohydrates. Paulsen and Todt (Hamburg) offer a review of the new and rapidly advancing subject of sugars containing nitrogen or sulfur as the hetero atom in the ring—a topic that presents new and difficult problems in nomenclature.

Greenwood and Milne (Edinburgh) present a discussion of starch enzymes, and Gorin and Spencer (Saskatchewan) discuss the structure of fungal polysaccharides. Shafizadeh (Montana) reviews recent activity in the study of the pyrolysis and combustion of cellulosic materials.

An obituary of Clifford Purves is written by Perlin, who has succeeded Purves as E. D. Eddy Professor at McGill University.

The editors note with regret the death of Dr. L. H. Cretcher, one of whose accomplishments in carbohydrate chemistry was the discovery of D-mannuronic acid in seaweed.

The Subject Index was prepared by Dr. L. T. Capell.

Columbus, Ohio	M. L. Wolfrom
Kensington, Maryland September, 1968	R. Stuart Tipson

1902-1965

Clifford B. Purves

24. Volume 24

Editors' Preface.—In the untimely death of Professor Melville L. Wolfrom on June 20, 1969, the world of carbohydrate chemistry lost a great and inspiring scientist, teacher, and writer. He was an Editor of *Advances in Carbohydrate Chemistry* from its inception in 1945 until his demise, except for 1950–1951 when he served on the Board of Advisors. On the death of Claude S. Hudson in 1952, he became the Editor, with the writer acting successively as Assistant Editor (1953–1954), Associate Editor (1955–1966), and Editor (1967 on). Upon assuming this responsibility, Professor Wolfrom made a firm resolve to continue to adhere to the high standards of scholarship and writing that had been insisted upon by Professor Hudson. The merits of this policy are attested to by the succession of laudatory reviews that every volume of this serial publication has earned. Because of this demonstrated success, it is the intention of the Editor that this policy shall be continued, even though, as Professor Wolfrom dryly remarked in the Preface to Volume 8, "the enforcement of such a policy is not without attendant difficulties."

In every volume of this serial publication, the nomenclature of carbohydrates has presented problems. In their solution, the leadership that from 1951 Professor Wolfrom so ably provided as Chairman of the Carbohydrate Nomenclature Committee (of the Division of Carbohydrate Chemistry of the American Chemical Society) was also of inestimable value in ensuring the use of correct nomenclature in all succeeding volumes of *Advances in Carbohydrate Chemistry*.

In September of 1968, Professor Wolfrom appointed Professor Derek Horton to the position of Assistant Editor, with the understanding that his association with this publication was to commence with Volume 24 (1969). This appointment was made with the intent of preserving continuity as regards the goals and standards already mentioned. At about the same time, he decided to amend the title of the publication in order to make clear that topics in biochemistry are, as indeed they always have been, subjects of discussion herein.

Finally, because of the resignation of Dr. R. C. Hockett from the Board of Advisors and the death of Professor S. Peat (a member of the Board of Advisors for the British Isles), he decided to appoint, as replacements, the scientists whose names now appear on the title page and to widen the scope of the latter Board by renaming it the Board of Advisors for the British Commonwealth.

Kensington, Maryland
Columbus, Ohio September, 1969

R. Stuart Tipson
Derek Horton

In this volume, each of the two long and authoritative articles started in Volume 23 is concluded: these contributions are by Isbell and Pigman (Washington and New York) on the mutarotation of sugars in solution and by Ball and Parrish (Natick) on sulfonic esters of carbohydrates.

In a masterly account of the nitro sugars, Baer (Ottawa) extends and updates Sowden's related chapter that appeared in Volume 6. Although only four years have elapsed since Ferrier (London) discussed the unsaturated sugars (Volume 20), progress in this field has been so great that he devotes an entire chapter to advances made in the interim.

Rees (Edinburgh), in a stimulating discussion of the formation of polysaccharide gels and networks, develops recent ideas on conformations of macromolecules that may well serve to direct the trend of future work in this area. Aspinall (Peterborough) provides a detailed insight into the structure of gums and mucilages concerning important correlations that have been made since the subject was examined in Volume 13.

Finally, two topics not hitherto covered are reviewed. Kiss (Basle) contributes a comprehensive summary of the present status of knowledge of the glycosphingolipids, carbohydrate compounds of considerable complexity and biological importance. McGale (Stoke-on-Trent) gives us a necessarily brief description of a little-explored group of substances, namely, the protein–carbohydrate compounds in human urine. An obituary of Richard Kuhn is provided by Baer (Ottawa), a former student of Kuhn's.

The Subject Index was compiled by Dr. L. T. Capell.

Kensington, Maryland R. Stuart Tipson
Columbus, Ohio November, 1969 Derek Horton

1900–1967

Richard Kuhn

25. Volume 25

Editors' Preface.—With this twenty-fifth Volume, *Advances in Carbohydrate Chemistry and Biochemistry* has completed its first quarter-century. This serial publication was initiated with the dual objective of presenting definitive accounts of the status of matured fields and of providing, for areas of high activity, critical evaluations that would serve as guidelines for future research. The past 25 years have seen an acceleration of research unprecedented in the history of science, and the extent to which *Advances* has usefully fulfilled a need, and yet provided flexibility in accommodating to change, may be judged by the frequency with which many of the older articles are still cited.

Over its lifetime, *Advances* has developed into a permanent source of reference in organized form for practically all of the major subdivisions of knowledge in the field of carbohydrates. It has also stimulated, by means of timely articles on active and controversial areas of research, the exploration of important fields that might otherwise have been neglected or investigated in a more haphazard fashion. The breadth of coverage, as originally conceived and subsequently maintained, has allowed the discussion of carbohydrates from the viewpoints of many specializations.

Structural and synthetic organic chemistry have certainly been highly influential, but biochemistry and physical chemistry have been no less important, and the techniques and ideas of agricultural chemistry, analytical chemistry, industrial chemistry and technology, microbiology, pharmacology, and many other disciplines have brought the full breadth of scientific inquiry to bear on this, the largest class of natural products.

In the present era, the idea of interdisciplinary research has become much in vogue, and it is therefore interesting to observe that the original Advisory Board for *Advances,* in setting a policy of studying a single major class of natural products by a broad spectrum of many classical disciplines, was instrumental in the achievement, over the years, of that very type of cooperation between specialists of different persuasions that is only now becoming properly recognized as an important trend in the future development of science.

Bearing in mind the not infrequent asseverations in the past by certain classical specialists (especially, organic chemists) that the study of carbohydrates is a narrow specialization, the tenacity of the Editors and the Advisory Board in adhering to the original interdisciplinary concept is a fitting testimony to the soundness of their ideas in the face of changing opinion.

The present volume was in the planning stage at the time of the death of Professor Melville L. Wolfrom, but it is one of the few volumes in the history of this Series not to have received his editorial attention. His influence always reflected the precision and accuracy that he applied in his own writings, and the present Editors will endeavor to meet these criteria.

This volume includes an obituary, contributed by J. R. Turvey, of the late Professor Stanley Peat, F.R.S., who was for many years a member of the Board of Advisors of *Advances,* and who served as Associate Editor for the British Isles for a number of years.

The separation of macromolecules by molecular-sieve techniques constitutes a major technical advance, especially for biochemists. The principles of the method and applications in the carbohydrate field are surveyed by S. C. Churms (Rondebosch).

The field of X-ray crystal-structure analysis is undergoing rapid evolution because of major advances in methodology; automatic diffractometers and computerized systems for data reduction have advanced the technique to the point where the solution of many simple structures is almost routine, and the chapter by G. Strahs (New York) surveys developments since the article by Jeffrey and Rosenstein in Volume 9 of *Advances.* The chapter marks a transition between the era of the classical crystallographer, who determined a structure for its own sake, and that of the newer generation of crystallographers concerned with the broader implications of a coordinated plan of attack, where crystallography provides the tool rather than the objective for the study of fine points of the molecular structure of carbohydrates in relation to their conformations and biological roles.

The wide-ranging subject of anhydrides of sugars and their derivatives is treated from the organic chemical viewpoint in three separate chapters in this volume. Because of the extensive literature on sugar anhydrides of various types, it was found impossible to treat developments in this whole area within the confines of one chapter, or even in the three chapters here presented; other aspects remain to be treated in future issues. Oxiranes (epoxides) are discussed by N. R. Williams (London), and ring-forming reactions, of aldoses, that result in the formation of 2,5-anhydro rings are delineated in the chapter by J. Defaye (Gif-sur-Yvette). The anhydrides of alditols are considered separately by S. Soltzberg (Delaware).

Modern methods of separation continue to reveal the complexity of mixtures of even simple carbohydrates in various natural sources, as demonstrated in the chapter by I. R. Siddiqui (Ottawa) on the sugars present in honey. The reactions of sugars with ammonia and amines, a field related to important problems in the food industry, constitute a subject of continuing interest, and are treated by M. J. Kort (Pietermaritzburg).

The polyfunctional nature of sugars can be exploited in the synthesis of a multitude of types of heterocyclic derivative, but the uninitiated reader may find the literature confusing and disorganized because of the plethora of structural types possible, even from simple reactions; H. El Khadem (Alexandria) has performed a valuable service by organizing the facts, fictions, and paradoxes in this domain.

In the final chapter, R. D. Marshall and A. Neuberger (London) explore the recent developments in our understanding of the structure and metabolism of glycoproteins, an area at the broad frontier of much advancing knowledge in modern biochemistry. The Subject Index was compiled by Dr. L. T. Capell.

A well-assorted, international representation of authorship is evident in recent volumes of *Advances*; the original British–American liaison on which the publication was founded has been substantially expanded to the international level. The present volume includes, in addition to contributions from North America and Great Britain, articles from continental Europe and, coincidentally, three separate chapters by authors based at different points on the African continent.

Kensington, Maryland R. Stuart Tipson
Columbus, Ohio November, 1970 Derek Horton

Stanley Peat
1902-1969

26. Volume 26

Editors' Preface.—In this volume, the twenty-sixth of *Advances in Carbohydrate Chemistry and Biochemistry*, is presented an article on conformational analysis of sugars and their derivatives by Durette and Horton (Columbus), who have brought into focus many of the recent advances in this field.

In a discussion of cyclic acyloxonium ions in carbohydrate chemistry, Paulsen (Hamburg) provides many striking illustrations of stable ions long considered only as transient reaction-intermediates. Their transformations furnish fascinating examples of the way in which the sugars can be used in furthering our basic knowledge of the stereochemical control of organic reactions.

Brady (Washington) offers an up-to-date account of cyclic acetals of ketoses that complements earlier articles on cyclic acetals of hexitols, pentitols, and tetritols (Barker and Bourne, Vol. 7) and of the aldoses and aldosides (de Belder, Vol. 20).

Butterworth and Hanessian (Montreal) have compiled useful tables of the properties of deoxy sugars and their simple derivatives; these tables are intended for use in conjunction with Hanessian's earlier chapter on deoxy sugars in Vol. 21. The biological elaboration of complex saccharides receives in-depth treatment in two articles in the present volume.

Shafizadeh and McGinnis (Missoula) have contributed an authoritative treatment of the morphology and biogenesis of cellulose and plant cell-walls, and a "companion piece" by Nikaido and Hassid (Berkeley) is a masterly account of the biosynthesis of saccharides from glycopyranosyl esters of nucleoside pyrophosphates.

An obituary of Melville L. Wolfrom has been written by Horton, a colleague of Wolfrom's at The Ohio State University. Because of the unusual significance of Wolfrom's contributions to carbohydrate chemistry, a list of his publications has been provided that will, we trust, prove useful to our readers.

The Subject Index was prepared by Dr. L. T. Capell.

Kensington, Maryland R. Stuart Tipson
Columbus, Ohio October, 1971 Derek Horton

1900–1969

Melville Lawrence Wolfrom

27. Volume 27

Editors' Preface.—The twenty-seventh volume in this serial publication presents the first part of a two-part article by Coxon (Washington) on the proton magnetic resonance spectroscopy of carbohydrates that emphasizes important recent advances and updates the article by Hall in Volume 19.

Moye (Sydney) provides a useful account of non-aqueous solvents for carbohydrates that brings together in one place a great deal of widely scattered information; his contribution should prove especially useful for the technologist, but the fundamental investigator will also find much valuable information from a seldom-stressed viewpoint.

The preparation, properties, and uses of sugars specifically labeled with isotopes of hydrogen are capably discussed by Barnett and Corina (Southampton); such compounds provide a most important tool for understanding organic and biochemical reaction-mechanisms of the sugars. Inch (Salisbury) delineates the use of carbohydrates in the synthesis and configurational assignments of optically active, non-carbohydrate compounds, some of which have powerful physiological effects as cholinergic and other drugs.

Zhdanov, Alexeev, and Alexeeva (Rostov-on-Don) contribute a comprehensive account of applications of the Wittig reaction in carbohydrate chemistry; this exceptionally versatile reaction will undoubtedly play a key role in much future synthetic work in the field.

It is, perhaps, not widely recognized that many enzymes, especially the hydrolases, possess a covalently linked carbohydrate moiety; in this Volume, Pazur and Aronson (University Park, Pennsylvania) bring into focus this aspect of enzyme structure by presenting a discussion of such glycoenzymes, that is, enzymes having a glycoprotein structure.

The obituary in this Volume, by Jamieson (Washington), pays tribute to William Werner Zorbach, a scientist remembered as an outstanding teacher and a gifted experimentalist in the chemistry of carbohydrates.

The Subject Index was prepared by Dr. L. T. Capell.

Kensington, Maryland
Columbus, Ohio October, 1972

R. Stuart Tipson
Derek Horton

1916–1970

William Werner Zorbach

28. Volume 28

Editors' Preface.—In a departure from the custom of annual publication that has been traditional for this Series, this twenty-eighth volume of *Advances* initiates a new schedule for publication of issues, accelerated to permit timely coverage of important topics in the face of greatly increased activity and interest on various fronts of the carbohydrate field. Appearance of this Volume in mid-1973, instead of the customary late autumn, reflects a decision to augment the scope and topicality of the Series by issuing volumes at intervals of about eight months instead of twelve.

This Volume continues a series of articles devoted to modem methods for the separation and characterization of carbohydrates; Part I of a Chapter by Dutton (Vancouver) that is concerned with the gas–liquid chromatography of sugars and their derivatives is presented. Although less than a decade has elapsed since Bishop's article on this subject appeared in this Series, the literature has since become vast and frequently confusing. The present article provides a critical analysis, together with extensive tabulated information, to guide the investigator in finding the best conditions for effecting separation of sugars and their derivatives.

Another Chapter of broad, general interest is the contribution of Feather and Harris (Columbia, Missouri, and Madison, Wisconsin) on the dehydration reactions of sugars. The complex, sequential transformations that occur in aqueous acid or alkali, and that proceed from initial, enediol derivatives through unstable, dicarbonyl intermediates to the myriad degradation products, are related to problems in many areas of carbohydrate technology on the one hand, and to such applications as colorimetric methods for analysis of carbohydrates on the other.

Deoxyhalogeno sugars have been of great synthetic interest in recent years, and Szarek (Kingston, Ontario) brings up to date the progress in this area since 1967, when the subject was last treated in this Series.

As a complement to the comprehensive Chapter on the biosynthesis of complex saccharides by Nikaido and Hassid in Vol. 26, where the products of biosynthesis were stressed, Kochetkov and Shibaev (Moscow) here emphasize the "nucleotide sugar" biosynthetic intermediates, themselves, in a detailed article on their chemistry and enzymology.

As part of a collection of articles focusing on individual enzymes acting on carbohydrates, this Volume features a contribution by Snaith and Levvy (Aberdeen) on α-D-mannosidase, a zinc-containing metalloenzyme.

The life and work of Laszlo Vargha, the Hungarian carbohydrate chemist, is the subject of an obituary article by Kuszmann (Budapest).

The Subject Index was compiled by Dr. L. T. Capell.

Kensington, Maryland   R. Stuart Tipson
Columbus, Ohio September, 1973   Derek Horton

1903–1971

Laszlo Vargha

29. Volume 29

Editors' Preface.—New developments in physical methodology for study of sugars, and expansion and consolidation of older ones, continue to attract the attention of researchers from widely differing areas. In this twenty-ninth volume, Hall (Vancouver) builds upon the background of his classic Chapter in Volume 19, together with the detailed treatment of modern proton magnetic resonance methodology presented by Coxon in Volume 27, to illustrate some of the specialized chemical and physical "tricks of the trade" for extracting useful information from complex, proton magnetic resonance spectra. With its emphasis on the pragmatic approach and with glimpses of the broad future potential of such tools as paramagnetic shift-reagents, Hall's article "Solutions to the Hidden-Resonance Problem" should appeal particularly to the investigator who is not a magnetic resonance specialist, but who seeks to exploit the analytical possibilities of proton magnetic resonance to their limit.

Practical applications are again stressed in the chapter by Lönngren and Svensson (Stockholm) on "Mass Spectrometry in Structural Analysis of Natural Carbohydrates." They build on the fundamentals of carbohydrate mass spectrometry, as laid down by Kochetkov and Chizhov in Volume 21, and demonstrate the profound analytical value of mass spectrometry for structural analysis of complex polysaccharides. In particular, this tool has dramatically increased the scope of the traditional methylation linkage-analysis procedure, especially when used in conjunction with gas–liquid chromatographic methods of separation. The latter topic is the subject of complementary Chapters by Dutton, one already published in Volume 28 and the other scheduled for publication in Volume 30.

The development of electrochemical procedures for industrial synthesis of alditols was initiated in the 1920s and the literature on electrochemistry of carbohydrates, from both synthetic and analytical viewpoints, is extensive. However, it is probably true to state that the average carbohydrate chemist has scant knowledge of the literature on preparative aspects of electrochemistry, or on polarography of sugars. The Chapter by Fedoroňko (Bratislava) should, therefore, fill an important need by integrating the work on the electrochemistry of carbohydrates that has matured during several decades.

Potential applications in industry also constitute an important aspect of the Chapter by Mizuno and Weiss (Shizuoka, Japan, and Worcester, Mass.) on the formose reaction. Although long known, this *in vitro* polymerization of formaldehyde to generate sugars gives rise to complex mixtures that have awaited the advent of modern separation methods to afford detailed understanding of the reaction and the individual products formed.

Biochemical aspects are featured in the Chapters by Kiss (Basle) on "β-Eliminative Degradation of Carbohydrates Containing Uronic Acid Residues" and by Kennedy (Birmingham) on "Chemically Reactive Derivatives of Polysaccharides." The β-eliminative degradation of uronic acid-containing sugar derivatives is an important consideration in structural work, especially on pectins, algal polysaccharides, and glycosaminoglycans.

The Chapter by Kennedy covers a very broad field, with emphasis on developments during the past decade on synthesis and applications of polysaccharides having functional molecules covalently attached. Such derivatives are of high current interest in the rapidly developing areas of immobilized enzymes and affinity chromatography, as well as in the more traditional applications of polysaccharide ethers and esters.

The obituary article by Goodman (Kingston, R. I.) describes the career of the late B. R. Baker, emphasizing especially Baker's contribution of useful synthetic procedures in the sugar field by application of neighboring-group reactions for configurational inversion.

The Subject Index was compiled by Dr. L. T. Capell.

The editors note with regret the passing of our friend H. G. Fletcher, Jr., on October 19, 1973.

Kensington, Maryland R. Stuart Tipson
Columbus, Ohio December, 1973 Derek Horton

Bernard Randall Baker

30. Volume 30

Editors' Preface.—The elucidation of structure of many complex, natural carbohydrates, especially polysaccharides and their conjugates, has progressed rapidly in the past few years, both in scope and precision, as a result of newer methods for specific degradation of these large molecules and for separating and identifying the products. In this thirtieth volume of *Advances,* Dutton (Vancouver) concludes his comprehensive treatment of published methodology for gas–liquid chromatography of sugars and their derivatives with the second part of a two-part Chapter; Part I was published in Volume 28. The present volume also features the first part of an article by Marshall (Miami) on the application of enzymic methods for structural analysis of polysaccharides.

Notable developments in the field of aminoglycosidic antibiotics have been recorded in recent years, especially by investigators in Japan. The medicinal scope of these therapeutic agents has been extended by useful semisynthetic variants, as well as by the discovery of new antibiotics produced microbially, and the rational design of effective, chemically modified agents has been guided by detailed biochemical investigations of the mode of action and of inactivation of these antibiotics. The outstanding team of Sumio Umezawa (Yokohama) and Hamao Umezawa (Tokyo) has been at the forefront of these advances, and these two investigators, respectively, contribute complementary Chapters on the structure and synthesis of the aminoglycosidic antibiotics, and on the biochemical mechanism of the development of resistance to them.

The disaccharide α,α-trehalose is the principal sugar of the circulatory system of insects, and its metabolism is discussed in a Chapter by Elbein (San Antonio). This subject has received relatively little study in comparison with the vast literature on metabolism of D-glucose in higher animals, but current interest in insect physiology and ecological problems associated with the use of traditional insecticides suggests that the time may be ripe for increased emphasis on the metabolism of α,α-trehalose. Sidebotham (London) contributes a survey of the extensive literature on the dextrans, microbial polysaccharides that are of considerable importance in several areas of applied biochemistry and technology, as well as being of fundamental interest.

This volume also presents the first of a projected, regular series of bibliographic articles that compile those carbohydrate structures that have been definitively established by crystallographic methods. Jeffrey (Pittsburgh) and Sundaralingam (Wisconsin) list here, with brief descriptive details, those crystallographically determined structures of sugars, nucleosides, and nucleotides that were reported during 1970–1972; future volumes will contain similar articles for successive calendar years.

The obituary article by Bacon (Aberdeen) and Manners (Edinburgh) provides an interesting insight into the personality and career of David J. Bell.

The Subject Index was compiled by L. T. Capell.

Kensington, Maryland R. Stuart Tipson
Columbus, Ohio June, 1974 Derek Horton

1905–1972

David J. Bell

31. Volume 31

Editors' Preface.—In this thirty-first volume of *Advances,* Williams (Swansea) surveys the deamination of carbohydrate amines and related compounds, updating earlier discussions by Peat (Vol. 2), Shafizadeh (Vol. 3), and Defaye (Vol. 25). Gelpi and Cadenas (Buenos Aires) provide a comprehensive treatment of the reaction of ammonia with acyl esters of carbohydrates; their article greatly extends that by Deulofeu (Vol. 4).

A chapter by Watson (Jackson, Miss.) and Orenstein (Boston, Mass.) brings the article by Hudson (Vol. 4) on the chemistry and biochemistry of apiose up to date. Lindberg, Lönngren, and Svensson (Stockholm) discuss the specific, chemical degradation of polysaccharides in an article that updates that by Bouveng and Lindberg (Vol. 15) and complements that by Marshall on their enzymic degradation (Vol. 30).

The extensive literature on the chemistry and interactions of seed galactomannans is surveyed by Dea and Morrison (Sharnbrook, England), thus adding to our previous articles on the chemistry of a variety of polysaccharides.

Glaudemans (Bethesda, Maryland) provides an interesting discussion on the interaction of homogeneous, murine myeloma immunoglobulins with polysaccharide antigens, and also describes the career of the late H. G. Fletcher, Jr.

In a continuation of our series of bibliographic articles on carbohydrate structures that have been ascertained by crystallographic methods, Jeffrey (Pittsburgh) and Sundaralingam (Madison, Wis.) treat those structures definitively established in 1973 and list all of those determined satisfactorily before 1970.

The Subject Index was compiled by L. T. Capell.

Kensington, Maryland	R. Stuart Tipson
Columbus, Ohio June, 1975	Derek Horton

1917–1973

Hewitt G. Fletcher

32. Volume 32

Editors' Preface.—In this volume, Wander and Horton (Columbus) provide a detailed discussion of the dithioacetals of sugars, compounds of considerable utility in the synthetic chemistry of carbohydrates because of the large number of reactions that are possible at the dithioacetal group and because of the hydroxyl groups of the acyclic sugar chain that are capable of selective protection or chemical transformation. The article complements that on the monothio derivatives of sugars by Horton and Hutson in Volume 18.

Neuberg's chapter in Volume 4 (1949) on biochemical reductions at the expense of sugars has been updated, as regards the utilization of sugars by yeasts, in an extensive article by Barnett (Norwich) that delineates the true complexity of the processes so frequently oversimplified through unjustified generalizations in standard biochemical texts.

Whistler, Bushway, and Singh (Lafayette), in collaboration with Nakahara and Tokuzen (Tokyo), have collected the information currently available on the sources, structures, and mode of action of noncytotoxic polysaccharides that display antitumor properties. This aspect of cancer chemotherapy has received scant attention elsewhere, and the chapter serves to bring together into common focus the extensive, but widely scattered, literature on the subject.

Dekker and Richards (Townsville) describe the occurrence, purification, properties, and mode of action of hemicellulases in an article that complements and extends our previous chapters on enzymes of interest to carbohydrate chemists. The present chapter treats, in fact, four major enzyme types, acting on L-arabinans, D-galactans, D-mannans, and D-xylans, from the standpoint of their action on the complex polysaccharide group known collectively as hemicelluloses.

Jeffrey (Upton) and Sundaralingam (Madison) continue our series of bibliographic articles, initiated in Volume 30, on carbohydrate structures by describing those that have been ascertained by crystallographic methods during 1974. As a new feature, they have introduced projection diagrams, produced by means of computer graphics, for depiction of the conformation of each of the organic molecules or ions.

The obituary article, by Ballou and Barker (Berkeley), describes the fascinating career of Zev Hassid, and particularly discusses his contributions to our knowledge of glycosyl esters of nucleoside pyrophosphates and the biosynthesis of polysaccharides.

The editors note with profound regret the passing on October 29, 1975, of our friend and erstwhile mentor Sir Edmund Hirst, a member of the Executive Committee of *Advances* from 1948 to 1950 and of our Board of Advisors from 1950 to 1952, an

Associate Editor for the British Isles from 1953 to 1954, a member of the Board of Advisors for the British Isles from 1955 to 1968, a member of the Board of Advisors for the British Commonwealth from 1969 to 1974 (Vol. 29), and a member of the Board of Advisors from 1974 (Vol. 30) until his death.

The Subject Index was compiled by Dr. L. T. Capell.

Kensington, Maryland R. Stuart Tipson
Columbus, Ohio February, 1976 Derek Horton

1899–1974

Zev Hassid

33. Volume 33

Editors' Preface.—In this thirty-third volume of *Advances,* Haines (Norwich) surveys the relative reactivities of hydroxyl groups in carbohydrates, as regards esterification, etherification, and other reactions, in an article that updates that by Sugihara (Vol. 8); the author has correlated a vast amount of widely scattered literature in a way that should prove particularly helpful to the synthetic chemist seeking routes to specific, partially substituted intermediates.

Also of particular interest to the synthetic chemist is the article by Hanessian and Pernet (Montreal); they have provided a comprehensive discussion of the synthesis of naturally occurring C-nucleosides, their analogs, and functionalized *C*-glycosyl precursors. Research on *C*-glycosyl compounds has made enormous strides since earlier articles were published by Haynes (Vols. 18 and 20), largely as a result of chemotherapeutic interest in such C-nucleosides as the formycin group.

The reactions of D-glucofuranurono-6,3-lactone are collated and discussed by Dax and Weidmann (Graz); this compound is of considerable importance as a precursor for synthesis of various, useful carbohydrates, but earlier volumes of *Advances* had not furnished the detailed focus that is developed here on the chemistry of this particular lactone.

Khan (Reading) has written a much-needed summary of the chemistry of sucrose, a subject that has burgeoned tremendously since its discussion by Levi and Purves in Vol. 4, largely as a result of efforts by the International Sugar Research Foundation to promote development of sucrose-based chemicals for potential, technological utilization. The comprehensive set of Tables of sucrose derivatives, included at the end of this Chapter, should prove particularly useful as a source of reference.

Larm (Uppsala) and Lindberg (Stockholm) educe comparative information on the structures of twenty pneumococcal polysaccharides, thus greatly extending the information previously presented by How, Brimacombe, and Stacey in Vol. 19. Improved techniques for polysaccharide structure-determination, largely pioneered in Lindberg's laboratory, have greatly improved the speed and reliability of structural characterization, to the point that serological classification of microorganisms on the basis of structures of their immunologically specific polysaccharides is becoming increasingly commonplace.

Rexova-Benkova and Marković (Bratislava) describe the action pattern and specificity, occurrence and formation, purification, and assay of pectic enzymes, a topic last discussed by Kertesz and McColloch in Vol. 5. Since that time, there have been important developments, notably in the discovery of enzymes that split glycosidic linkages by β-elimination.

Complementing our continuing series of bibliographic articles on the structures of simple carbohydrates as ascertained by X-ray crystallographic methods (Jeffrey and Sundaralingam, Vols. 30–32), Marchessault and Sundararajan (Montreal) now provide a similar bibliography of crystal structures of polysaccharides determined during 1967–1974, thus updating the information in the article by Marchessault and Sarko in Vol. 22.

An informative obituary describing the life and work of Alfred Gottschalk has been contributed by Neuberger (London).

The editors note with regret the passing of our friend Edward J. Bourne on November 30th, 1974.

The Subject Index was compiled by Dr. L. T. Capell.

Kensington, Maryland R. Stuart Tipson
Columbus, Ohio June, 1976 Derek Horton

1894-1973

Alfred Gottschalk

34. Volume 34

Editors' Preface.—In this volume of *Advances,* Černý and Staněk, Jr. (Prague) contribute a comprehensive article on the 1,6-anhydro derivatives of aldohexoses. This class of anhydro sugars, earlier termed the hexosans, constitutes by far the largest class of sugar anhydrides to have been studied. The scope of the topic is now so extensive that the Editors, in view of their earlier practice of limiting chapter size in individual volumes of *Advances* in the interest of diversification of subject material, were inclined to divide this chapter into two parts, to appear in successive volumes. However, as the latter practice has elicited some unfavorable reaction from readers on previous occasions, it was decided to present the article here in its entirety. The chapter complements previous ones on other classes of anhydro derivatives, in particular, recent ones on anhydrides of the oxirane (epoxide) and 2,5-anhydro (oxolane) types.

Although the cyclic acetals of aldoses and aldosides were the subject of an article by De Belder (Uppsala) as recently as Volume 20 (1965), the subject has been in a state of particularly active development in recent years, because of the exceptional synthetic utility of these derivatives and as a result of access to other types of cyclic acetals through the advent of new preparative methods and improvements in methods for the analysis of reaction mixtures. To keep the subject up to date, De Belder has now contributed a new article that highlights the literature since his chapter in Volume 20, and he also provides a supplement to the tables of data featured with the original article.

The Koenigs–Knorr reaction is one of the oldest reactions in the carbohydrate literature, yet the rational use of this synthesis for generating the mixed acetal group that links sugar residues in glycosides and oligosaccharides remains enigmatic and capricious; generalizations are difficult to make and the reaction continues to challenge modem investigators. In this issue, Igarashi (Osaka) has provided a detailed comparison of numerous efforts in this area, with particular emphasis on the relationship between anomeric ratios in the products and the conditions used for the reaction.

With the advent of new, commercial procedures for production of D-fructose and high-fructose sweeteners, profound changes are taking place in the traditional sweetener industry, and human diet-patterns face substantial changes toward an increased, direct intake of D-fructose. The article by Chen and Whistler (Purdue) on metabolism of D-fructose is thus particularly timely, as it brings into detailed focus the frequently overlooked metabolic differences between the conventional dietary sugars and this ketose acting alone.

The 1975 literature of crystal structures of carbohydrates, nucleosides, and nucleotides has been summarized by Jeffrey (Upton, N.Y.) and Sundaralingam (Madison); following the procedure used by them in previous such bibliographies, the depictions have been re-drawn by means of computer graphics so as to present the structures in a format that is most readily comprehensible for organic chemists and biochemists. Corrections in the original data have been made where necessary, especially as regards specification of the correct enantiomer.

The untimely death of our friend and colleague Edward J. Bourne was noted briefly in the previous volume; in this issue, Weigel (Egham) contributes a sensitive and personal account of Bourne's life and scientific work.

The editors note with regret the passing of our friends Sir Edmund Hirst on October 29, 1975, J. K. N. Jones on April 13, 1977, and W. W. Pigman on September 30, 1977.

The Subject Index was compiled by Dr. L. T. Capell.

Kensington, Maryland　　　　　　　　　　　　　　　　　　　　　　R. Stuart Tipson
Columbus, Ohio September, 1977　　　　　　　　　　　　　　　　　Derek Horton

1922–1974

Edward J. Bourne

35. Volume 35

Editors' Preface.—In 1961, Ferrier (Wellington, N. Z.) began a study of the condensation of phenylboronic acid with the diol groupings of various glycosides, and since then, as a result of research by him and many other workers, the subject has developed rapidly and has afforded information that is of great potential value to synthetic carbohydrate chemists; Ferrier now provides us with a fascinating account of the progress made to date in the study of carbohydrate boronates. This unifying picture of the use of these cyclic, protected sugar derivatives should afford considerable help to the chemist searching for novel approaches in synthesis by use of these versatile protecting groups.

In an article that focuses particularly on some of the more-recent developments, Grisebach (Freiburg im Breisgau) discusses the biosynthesis of sugar components of antibiotic substances, a field that has shown major advances since the time of the article by Dutcher in Volume 18 of *Advances*. The chapter constitutes an integrating complement to the articles on aminocyclitol antibiotics by the Umezawas (Volume 30).

Goldstein (Ann Arbor, Michigan) and Hayes (Madison) contribute a monumental chapter on lectins, the specific carbohydrate-binding proteins present in both plant and animal species. The remarkable specificity of certain carbohydrate–protein interactions has far-reaching implications in biochemistry, the full significance of which is only just beginning to be properly understood. As this subject has not previously been treated in depth in *Advances*, these authors have written a comprehensive history of the subject, starting with Stillmark's discovery of plant agglutinins in 1888, and proceeding to 1977; this article brings together in one place an enormous amount of information scattered throughout the literature and should constitute the definitive treatment of lectins for many years to come.

Another biochemical topic, the biochemistry of plant galactomannans, is discussed by Dey (Egham, Surrey); the article rounds out aspects of the field that are complementary to those treated by Gorin and Spencer in Volume 23 (on fungal polysaccharides) and Dea and Morrison in Volume 31 (on the chemistry and interactions of seed galactomannans).

In a continuation of the series of bibliographic articles on the structures of polysaccharides as established by X-ray crystallographic methods, Sundararajan (Mississauga, Ontario) and Marchessault (Montreal) present the information recorded in the literature during 1975, thus updating their article in Volume 33. Since the latter was written, SI units have become generally adopted; the time-honored Ångstrom unit (Å) has now fallen into disuse, and so it will no longer be employed in this Series.

The death of our friend and mentor Sir Edmund Hirst was briefly noted in the Preface to Volume 32. A full account of his career and wide-ranging achievements is given here by Stacey (Birmingham) and Manners (Edinburgh).

The Subject Index was compiled by Dr. L. T. Capell.

Kensington, Maryland
Columbus, Ohio February, 1978

R. Stuart Tipson
Derek Horton

1898 – 1975

Sir Edmund L. Hirst

36. Volume 36

Editors' Preface.—The central role of carbohydrates in nutrition has long supported a broad base in technology and economics in relation to the two most abundant dietary carbohydrates, namely sucrose and starch. The utilization of starch as a basis for preparing nutritive sweeteners has been established since the 19th century. Although the commercial production of D-glucose and of malto-oligosaccharides by acid hydrolysis of starch dates back many years, the starch industry has recently pioneered the successful exploitation of immobilized enzymes for effecting this process, and furthermore, the classic reaction for isomerization of D-glucose to D-fructose, long sought as a preparative route for preparing D-fructose as a high-sweetness sugar to compete with invert sugar, has also benefited from innovative technology utilizing immobilized enzymes. MacAllister (Clinton, Iowa) provides a very informative account of the development of these processes and their introduction on the industrial scale, showing how starch technologists have brought hydrolytic and isomerizing enzymes out of the laboratory and into successful commercial utilization.

Although the glycosides of D-glucuronic acid are well recognized as being the result of a major metabolic pathway for detoxification and excretion of unwanted toxic substances in mammalian metabolism, the volume and diversity of literature on these glycosides may be difficult to assimilate by the average chemist. The subject has been treated in early volumes in this series, and in several books, but in this volume, Keglević (Zagreb) has emphasized the chemical aspects and more recent developments in what should prove a very useful and comprehensive account of the subject.

As a complement to earlier articles in this Series on the chemistry of nucleosides and nucleotides, Ikehara, Ohtsuka, and Markham (Osaka) provide here a survey of methodology for the synthesis of polynucleotides that has evolved rapidly in recent years, notably as a result of the ingenious work of Khorana and his school. In devising specific reactions for linkage of the carbohydrate residues in nucleotide sequences through phosphoric diester bonds, there are particular differences in methodology required according to whether the sugar components are D-ribofuranosyl or 2-deoxy-D-*erythro*-pentofuranosyl. This area has seen the development of considerable finesse in the use of novel protecting agents in the carbohydrate portion, to provide compatibility with other structural components in systems of considerable complexity. This Chapter should provide a source of reference and inspiration not only to researchers engaged in nucleotide synthesis but also to chemists involved in many other synthetic endeavors that require carefully selected choice of suitable protecting groups in carbohydrate systems.

The Chapter by Wilkie (Aberdeen) on the hemicelluloses of the Gramineae serves to illustrate that not all areas of complex carbohydrates are yet amenable to neat and clear-cut structural interpretations as a result of modern technology. The article emphasizes the need for continued caution in attribution of precise structures to this enigmatic class of polysaccharides, and final answers cannot yet be written on their structures and roles.

The pioneering work of Dr. Allene Jeanes and her coworkers on the extracellular polysaccharides of micro-organisms is familiar to most researchers through the familiar NRRL system for classifying microbial strains, but the potential of certain of these polysaccharides, notably xanthan, in technology and in food products has only become widely recognized and developed in recent years. The article by Sandford (Peoria), one of Dr. Jeanes's former collaborators, provides the reader with a very comprehensive survey of the status of this rapidly growing field.

Sundararajan and Marchessault (Montreal) furnish here a continuing article in the series of bibliographic surveys of crystal-structure work on polysaccharides; the literature from 1976 illustrates the continuing effort still required in the understanding of even such a seemingly simple polysaccharide as cellulose.

Two chemists, perhaps not so widely known as their work might have deserved, but who made notable contributions to the carbohydrate field from rather widely different viewpoints, are the subjects of biographical memoirs in this volume. The story of the life and work of John A. Mills is provided by S. J. Angyal (Sydney), and that of Joseph V. Karabinos by W. W. Binkley (Kent, Ohio).

Kensington, Maryland	R. Stuart Tipson
Columbus, Ohio December, 1978	Derek Horton

John A. Mills
1919–1977

1920–1977

Joseph V. Karabinos

37. Volume 37

Editors' Preface.—Almost two decades ago, Phillips discussed, in Volume 16 of this series, the behavior of carbohydrates when subjected to ionizing radiation. Since that time, there has been in this area extensive new work, fundamental and applied, encompassing both simple and complex carbohydrates. Development in this field has been greatly aided by improved analytical methodology, and von Sonntag here provides a broad and comprehensive account of the more recent developments.

A nutritionally and commercially important vitamin, L-ascorbic acid, was the subject of a classic article by Fred Smith in Volume 2 (1946) of *Advances*, but since that time there has been an enormous expenditure of research and development effort directed toward improving the practical synthesis of this unusual sugar, and this is set out in a comprehensive treatment by the Crawfords in this volume. In addition to the direct relevance of this work to the synthesis of L-ascorbic acid, the chapter assembles a great body of synthetic "know-how" of broad general interest for the synthetic chemist, and it also illustrates the value of alternative or complementary microbiological procedures in synthesis.

In the first of a projected series of related articles on the subject of glycoproteins, Montreuil contributes a compact·but far-reaching chapter that sets the stage of modem structural work on the glycan portions of these molecules. Although these studies had their genesis in painstaking, early investigations on the oligosaccharides of milk, only recently has the wide-reaching significance of glycoproteins as biological information-carriers become fully recognized. The explosive growth of research in this area has arisen from this awareness, coupled with the development of separation and structure-determination methodologies of a finesse scarcely conceivable only a decade ago.

The covalent, chemical attachment of sugars or glycans to proteins is not a new concept, but interest in such synthetic structures as glycoprotein analogs ("neoglycoproteins") has developed rapidly in recent years as a result of advances in our knowledge of glycoprotein structure and function, and the need for reference analogs of defined carbohydrate composition. In this volume, Stowell and Lee provide a comparative and critical treatment of these synthetic, carbohydrate–protein conjugates, and survey their applications as synthetic antigens and lectin substrates, and in metabolic studies.

The chapter by Dey focuses on the multitude of oligosaccharides and polysaccharides containing α-D-galactosidic linkages that are found in the plant kingdom. Although the sucrose-related family of oligosaccharides, as exemplified by raffinose and stachyose, are very well known and were discussed a quarter of a century ago by French in Volume 9, Dey develops in this volume the broader range of products formed by α-D-galactosylation of numerous acceptors; these products have important implications in the metabolism and classification of all higher plants, and in nutritional considerations with various plant foodstuffs.

In continuation of established procedure, this volume includes a bibliography, contributed by Jeffrey and Sundaralingam, of crystal structures, published during 1976, for sugars, nucleosides, and nucleotides; the graphic depictions were computer-generated from the original coordinates and are oriented to the viewpoint familiar to carbohydrate chemists, and any errors found in the original reports have been corrected.

W. Ward Pigman, one of the founders of this series, co-editor of the first four volumes, long-time member of the Advisory Board (1950–1977), and close friend of the present Editors, passed away in 1977, as briefly noted in the Preface to Volume 34. His friend and colleague, Anthony Herp, here contributes a detailed account of Pigman's life and scientific career.

The Subject Index was compiled by Dr. L. T. Capell.

Kensington, Maryland
Columbus, Ohio December, 1979

R. Stuart Tipson
Derek Horton

W. Ward Pigman

38. Volume 38

Editors' Preface.—In Volume 38, the thirtieth in this series that has been edited by the Senior Editor, a discussion of the ^{13}C-nuclear magnetic resonance spectroscopy of polysaccharides is provided by P. A. J. Gorin (Saskatoon). This technique has opened up a major, new avenue for determination of the structure of these macromolecular carbohydrates, and has proved much more fruitful than ^{1}H-n.m.r. spectroscopy, a methodology discussed largely in the context of simple sugars, by B. Coxon in Volume 27 and L. D. Hall in Volume 29.

R. W. Binkley (Cleveland) complements and updates the article by G. O. Phillips in Volume 18 on the photochemical reactions of carbohydrates, with particular emphasis on organic-chemical aspects. A comprehensive treatment of fluorinated carbohydrates is contributed by A. A. E. Penglis (Oxford); this is a subject touched upon briefly in previous articles (by L. J. Haynes and F. H. Newth in Vol. 10; T. G. Bonner, F. Micheel, and A. Klemer in Vol. 16; J. E. G. Barnett in Vol. 22; and H. Paulsen in Vol. 26).

A review of the synthesis and reactions of the gulono-1,4-lactones and related derivatives is provided by T. C. Crawford (Groton). F. M. Unger (Vienna) contributes an article on the biological significance of 3-deoxy-D-*manno*-2-octulosonic acid (Kdo) and the chemical approaches for synthesis of this important compound. J. Järnefelt and coworkers (Helsinki) have written a detailed description of the use of the time-honored methylation technique, as adapted for modem, micro-scale work in the structural analysis of glycoproteins by J. Montreuil in Volume 37.

A continuing article in the series of bibliographic surveys of crystal-structure work on carbohydrates, nucleosides, and nucleotides, covering the literature thereon for 1977–1978, is provided by G. A. Jeffrey (Pittsburgh) and M. Sundaralingam (Madison); these surveys not only present crystallographic results in a pictorial format readily comprehended by chemists and biochemists, but they correct errors found in the original-literature interpretations.

H. S. El Khadem (Houghton) has written an interesting account of the life and work of Emil Hardegger, a Swiss sugar chemist whose work has been, perhaps, less well recognized than it might have been; a useful feature of this article is the citation of unpublished research that is available in doctoral dissertations.

Kensington, Maryland R. Stuart Tipson
Columbus, Ohio June, 1980 Derek Horton

1913 – 1978

Emil Hardegger

39. Volume 39

Editors' Preface.—A subject of continuing interest to carbohydrate chemists, namely, the selective removal of protecting groups, is here discussed by A. H. Haines (Norwich) in an article that updates and extends those of the same author in Vol. 33 and of J. M. Sugihara in Vol. 8, which were devoted more to the selective introduction of protecting groups. Of such protecting groups, the cyclic acetals, particularly 1,3-dioxolanes and 1,3-dioxanes, have played a major role since 1895, when Emil Fischer first described such derivatives of glycoses; and articles by A. N. de Belder, in Vols. 20 and 34 (on the cyclic acetals of the aldoses and aldosides), by R. F. Brady, Jr., in Vol. 26 (on those of the ketoses), and by S. A. Barker and E. J. Bourne in Vol. 7 (on the acetals of tetritols, pentitols, and hexitols) have discussed their synthesis and utility as Intermediates. However, in this volume, J. Gelas (Clermont-Ferrand) describes an entirely different aspect of their chemistry in which the cyclic acetal groupings themselves serve as functional groups in synthetic transformations.

A subject of both academic and industrial importance is the synthesis and polymerization of anhydro sugars, and C. Schuerch (Syracuse) contributes an extremely valuable account of the breakthroughs achieved by German and Latvian chemists, and in his own laboratory, in the application of cationic polymerization in this field. He also brings articles on the synthesis of anhydro sugars, by S. Peat (Vol. 2), R. J. Dimler (Vol. 7), N. R. Williams (Vol. 25), J. Defaye (Vol. 25), and M. Černý and J. Staněk, Jr. (Vol. 34), up to date.

An article by R. Khan (Reading) constitutes an all-encompassing treatment of the chemistry of maltose, a disaccharide whose industrial utility as a raw material has yet to be realized.

Both enantiomers of fucose are widespread in nature, and developments in their chemistry and their metabolism and biochemistry are treated by H. M. Flowers (Rehovot).

In Vol. 32, J. A. Barnett (Norwich) surveyed the utilization of the monosaccharide components of the common glycosides by yeasts, and he now extends this to an examination of their utilization of disaccharides.

Some attention had been devoted by J. F. Kennedy (Vol. 29) to affinity-chromatography matrices derived from polysaccharides, and the subject is discussed extensively in this volume by J. H. Pazur (University Park).

A highly interesting obituary article by C. E. Ballou (Berkeley) describes the career of K. P. (G.) Link and his contributions to carbohydrate chemistry.

The Subject Index was compiled by Dr. L. T. Capell.

Kensington, Maryland	R. Stuart Tipson
Columbus, Ohio March, 1981	Derek Horton

1901–1978

Karl Paul (G.) Link

40. Volume 40

Editors' Preface.—Although the many aspects of the biochemistry and biological properties of most of the naturally occurring monosaccharides have received extensive study, corresponding examination of their enantiomers has been sparse because of their commercial unavailability to date. However, with the advent of improved methods for resolving an enantiomeric mixture into the pure D and pure L form, the time was ripe for a detailed treatment of the chemical synthesis of such DL mixtures from non-carbohydrate sources. This topic constitutes the main focus of the article herein by Zamojski, Banaszek, and Grynkiewicz (Warsaw), which greatly extends previous treatments of the subject that were provided by Lespieau (Vol. 2), Mizuno and Weiss (Vol. 29), and Černý and Staněk (Vol. 34). These advances will undoubtedly lead to eventual, commercial availability of the "unnatural" enantiomers of the monosaccharides, and their intensive study by biochemists (who will, at last, find it imperative to state whether the D or the L form of a sugar was employed in their investigation); the possibility of applicability in the food industry may also be envisaged, and these sugars may have exciting potentialities in pharmaceutical chemistry and medical use.

Schauer (Kiel) provides a detailed discussion of the chemistry, metabolism, and functions of sialic acids, a subject first dealt with by Zilliken and Whitehouse in Volume 13 of this Series, when our knowledge of these acids was still in its formative stages, but great strides have since been made as a result of the introduction of such techniques as nuclear magnetic resonance spectroscopy, gas–liquid chromatography, and mass spectrometry. In the course of this article on the sialic acids, a term that encompasses all N-and O-acylated derivatives of neuraminic acid, it becomes apparent that the latter name, introduced by Klenk in 1941, was a most unfortunate choice as regards naming its derivatives, whereas the name "neuraminulosonic acid" would have lent itself to ready naming thereof. Furthermore, it points up the inadequacy of Rule Carb-9 of the *IUPAC-IUB Tentative Rules for Carbohydrate Nomenclature (1969)* as a guide in naming the conformers of such compounds by use of the *IUPAC-IUB Conformational Nomenclature*, whereas the British–American Rules (1963) are readily applicable, and provide intelligible names.

An article by Li and Li (Tulane University, LA) on the biosynthesis and catabolism of glycosphingolipids serves to extend that by Kiss (Vol. 24), which dealt mainly with the chemistry of these compounds. Schwarz and Datema (Giessen) provide a detailed account of the lipid pathway of protein glycosylation and of its inhibitors, and then discuss the biological significance of protein-bound carbohydrates, thereby affording a companion chapter to that by Montreuil (Vol. 37) on the primary structure of glycoprotein glycans.

Finally, Sundararajan and Marchessault (Mississauga, Ontario) bring up to the year 1979 their previous articles that provide a bibliography of crystal structures of polysaccharides as established by X-ray crystallographic and electron-diffraction methods.

The Subject Index was compiled by Dr. L. T. Capell.

Kensington, Maryland   R. Stuart Tipson
Columbus, Ohio August, 1982   Derek Horton

41. Volume 41

Editors' Preface.—In perhaps no other field of biological chemistry has n.m.r. spectroscopy played such an important role as it has in the structural investigation of the carbohydrates. Its use as an investigative tool has had significant implications across the whole range of carbohydrates, from simple sugar derivatives to complex polysaccharides and glycoconjugates. It is therefore fitting that, in *Advances*, this technique should constitute a sustained theme of interconnected articles that are devoted to various important groups of carbohydrates and to the implications of rapid advances in instrumental methodology.

As well demonstrated by Gorin's article in Volume 38 of this Series, carbon-13 n.m.r. spectroscopy has proved to be of profound significance in the structural investigation of polysaccharides; it has taken its place in complementing modern versions of such traditional techniques as methylation analysis and periodate oxidation, and may in large measure replace them as our library of reliable reference data for the simple sugar constituents is consolidated.

In the present volume, a most significant step in this direction is taken by Klaus Bock and Christian Pedersen (Lyngby, Denmark) in their extensive and careful compilation of carbon-13 data for a wide range of monosaccharides and their derivatives. The data are conveniently arranged in a selection of representative Tables, and the fact that the authors have themselves conducted extensive verification of the data presented offers the user a measure of convenience and confidence that could never be met by the scattered and often conflicting data in the primary literature.

In a similar vein, but with reference to proton-n.m.r. spectroscopy, Vliegenthart and coworkers (Utrecht, The Netherlands) have assembled the fruits of their detailed, comparative studies, by state-of-the-art, n.m.r. instrumentation, on a large number of carbohydrates related to glycoproteins. Much of this work was conducted with materials isolated in the laboratories of J. Montreuil (Lille, France), who contributed a landmark article on the structure of glycoproteins to Volume 37 of *Advances*. The present, complementary article displays the great power of high-field n.m.r. spectroscopy in applications related to glyco-conjugates of considerable complexity.

An article by Barreto-Bergter (Rio de Janeiro, Brazil) and Gorin (Saskatoon, Canada) likewise invokes strong emphasis on n.m.r. methods for structure determination, in this instance, by use of carbon-13 techniques in delineating the structural chemistry of polysaccharides from fungi and lichens.

In comparison with the foregoing complex polysaccharides and conjugates, the structure of the world's most abundant chemical compound, namely, cellulose, may seem prosaic indeed, and yet it is quite astonishing that, despite a high level of sophistication in our understanding of the mode of biosynthesis of many rare, complex carbohydrates, we still have remarkably little definitive knowledge of the way in which Nature builds this ubiquitous, plant polysaccharide. In an article that offers challenges to established dogmas and invites fresh thought, Deborah P. Delmer (now of Dublin, California) challenges the validity of conclusions that have often been taken for granted and emphasizes the need for open-minded, new research on the biosynthesis of the cellulose fiber.

The dramatic success of antibiotics for therapeutic control of many microbial infections has tended to overshadow the value of immunochemical approaches. The article by Jennings (Ottawa, Canada) provides an up-to-date discussion of the structures of a variety of bacterial capsular polysaccharides and serves to emphasize the important uses that such compounds have as human vaccines.

In a volume having strong emphasis on polysaccharide topics, it is especially appropriate to recognize the life and work of J. K. N. Jones. The article here contributed by Szarek and Hay (Kingston, Ontario, Canada) and Stacey (Birmingham, England) provides a sensitive account of Jones's work on both sides of the Atlantic, and includes a useful appendix that lists his scientific publications.

The Editors note with regret the recent passing of Louis Malaprade, University of Nancy, discoverer of the stoichiometric oxidation of glycols by periodate, a reaction that has had such profound implications in the structural investigation of carbohydrates; and of Karl Freudenberg, Heidelberg, last surviving student of Emil Fischer's, pioneer of important stereochemical concepts, and a scientist whose extensive contributions to synthesis included the classic, widely used acetone derivatives (isopropylidene acetals) of the monosaccharides.

The Subject Index was compiled by Dr. Leonard T. Capell.

Kensington, Maryland
Columbus, Ohio April, 1983

R. Stuart Tipson
Derek Horton

1912 – 1977

John Kenyon Netherton Jones

42. Volume 42

Editors' Preface.—In this volume, S. J. Angyal (Kensington, Australia) discusses the use of ^{1}H-n.m.r. spectroscopy in determining the composition of reducing sugars in solution. Applications of the n.m.r. technique have greatly advanced our understanding of the tautomeric behavior of sugars. Angyal has himself contributed much in this field, and his article complements from a modern perspective the information, deduced largely from such classical methods as polarimetry, provided by H. S. Isbell and W. W. Pigman in Vols. 23 and 24.

Branched-chain sugars were largely a curiosity when the natural occurrence of those then known was treated by F. Shafizadeh in Vol. 11, but the great variety of these actually now shown to exist has stimulated intense efforts by organic chemists to develop methods for their synthesis. J. Yoshimura (Yokohama, Japan) here treats in depth the application of a wide range of synthetic procedures for the generation of specific branching in sugar structures.

Sugar analogs having atoms other than oxygen in the hemiacetal ring have become of high interest from the standpoint of synthetic challenge and for their biochemical implications. H. Yamamoto and S. Inokawa (Okayama, Japan) introduce to this Series recent work directed toward such analogs having phosphorus as the ring atom.

K. Bock and C. and H. Pedersen extend the article in Vol. 41 by the first two authors, on the ^{13}C-n.m.r. spectroscopy of monosaccharides, to a compilation of such data for oligosaccharides that should prove of great value as a source of reference.

K. Antonakis (Villejuif, France) presents a discussion of ketonucleosides, compounds of interest in synthesis and in biological roles; they have not hitherto been comprehensively examined in this Series.

The nature of the plant cell-wall is still surprisingly little understood; P.M. Dey and K. Brinson (Egham, England) bring into current perspective an article thereon by Shafizadeh and McGinnis in Vol. 26. As part of a continued series of articles on classes of enzymes acting on carbohydrates, A. Kaji (Takamatsu, Japan) here discusses L-arabinosidases, thus adding to the detailed treatment of other such enzymes in earlier volumes (β-L-glucosiduronase, by G. A. Levvy and C. A. Marsh in Vol. 14; α- and β-D-galactosidases, by K. Wallenfels and O. P. Malhotra in Vol. 16; and α-D-mannosidase, by S. M. Snaith and G. A. Levvy in Vol. 28).

The life and work of Dexter French, who contributed so much to our knowledge of starch, is sensitively treated by his student J. H. Pazur (University Park, Pennsylvania). The pioneering discovery by French and Rundle, in the carbohydrate field, of a helical biopolymer in complexes of amylose predates the widely celebrated work with proteins and nucleic acids where the concept of a helical conformation revolutionized our understanding of the structure and function of these natural macromolecules.

The Editors note with regret the deaths of an unusually large number of well known carbohydrate chemists, including Konoshin Onodera, Leslie F. Wiggins, and Fred Shafizadeh.

Kensington, Maryland R. Stuart Tipson
Columbus, Ohio May, 1984 Derek Horton

1918–1981

Dexter French

43. Volume 43

Editors' Preface.—The make-up of this volume of *Advances* deviates somewhat from the accustomed format, even though the major themes of advances in carbohydrate structural methodology, synthetic methods related to important biomolecules, and structural characterization of key natural carbohydrates are all represented.

Two of the chapters are focused on the linkage between a glycosyl group and an oxygen or nitrogen atom of amino acids and proteins. The chapter by Dill (Clemson), Berman (San Francisco), and Pavia (Avignon, France) details the structural-characterization aspect of these linkages as it may be provided by carbon-13 nuclear magnetic resonance spectroscopy. This chapter supplements other recent chapters on practical aspects of ^{13}C-n.m.r. spectroscopy, a technique that has become established as a prime tool for structure elucidation in the field of complex carbohydrates and their conjugates. These authors illustrate the great value of modern instrumentation for determination of the structure of glycopeptides in relation to the entire glycoproteins of which they constitute fragments.

The chapter by Dill and coworkers is complemented by that of Garg and Jeanloz (Boston) on chemically synthesized glycosyl derivatives of sugar–amino acid conjugates in which sugar residues are attached to residues of L-asparagine, L-serine, and L-threonine. In addition to treating the various chemical synthetic approaches that have been used to generate these sugar–amino acid linkages, the chapter discusses the chemistry of this linkage in reactions brought about by acids, bases, and enzymes, and also includes an extensive tabulation of compounds of this type that have been prepared.

Casu (Milan, Italy) contributes a landmark article on the structure and biological activity of heparin. Although this subject was reviewed by Foster and Huggard in Volume 10 (1955) of *Advances,* enormous strides have been made in very recent years in understanding this enigmatic glycosaminoglycan that has so long played an important role in anticoagulant therapy. Only now are we beginning to understand the exquisite complexity of this molecule and its remarkable and varied role in a host of interrelated, biochemical processes in the circulatory system. In his article, Casu has brought together in a thoroughly integrated manner our current understanding of the chemical structure of heparin, its physical constitution, and its multiplicity of biochemical functions, while at the same time clearly pointing out the need for considerable further work before we can hope to approach a full understanding of this remarkable biomolecule.

In 1964 (Vol. 19), Jeffrey and Rosenstein (Pittsburgh) and in 1970 (Vol. 25), Strahs (N.Y.C.) provided articles on crystal-structure analysis in carbohydrate chemistry, but then, in 1974 (Vol. 30), this series initiated as a regular feature a bibliography of crystal structures of carbohydrates, nucleosides, and nucleotides. At the outset thereof, there was little coordination between the definitive, three-dimensional structures established by X-ray crystallographers on the one hand, and understanding of the true three-dimensional shapes of carbohydrates as they apply to the interpretation of organic-chemical and biochemical processes. One reason for this problem was the fact that little of the literature of crystallography was accessible, or even comprehensible, to the average biochemist or organic chemist, and there was an assumption on their part that a crystallographic analysis provided the ultimate and irrefutable proof of structure, notwithstanding the fact that crystallographers were frequently quite cavalier in the enantiomeric representation of structures, so that the crystallographic literature was replete with confusing articles purporting to give definitive structures that misleadingly depicted the wrong enantiomer; moreover, there would often be numerical errors in the parameters reported.

In the series of articles presented from 1974 (Volume 30) until now by Jeffrey (Pittsburgh) and Sundaralingam (Madison), a major effort has been made to bridge this gap by providing brief interpretations of crystallographic-structure studies that would bring out the essential content of importance to structural chemists and biochemists, and present, from the original crystal data, a structural depiction that would provide an accurate illustration of the molecule drafted according to conventions long familiar to organic chemists and biochemists. This series has been extremely successful, and has led to extensive cross-fertilization between the fields, with the result that organic chemists and biochemists are now increasingly depicting their molecules in representations that are close to the real natural shape, and crystallographers have become increasingly aware of the need for providing depictions that help to answer the structural questions of chemists and biochemists in general, and to address the question of correct enantiomorphism with due care. In the process of compiling these summaries, the opportunity has been taken to assess the validity of all numerical data, and, if necessary, to make corrections.

Now that automated diffractometers and sophisticated computer data-reduction systems have become routinely used, the task involved in conducting an X-ray crystal-structure examination has been enormously simplified. As a result, the number of structures determined each year by crystallography has escalated dramatically. The logistics of maintaining this series in the original form would now entail such stupendous demands on our space that a volume of *Advances* could be completely occupied by this one article alone. As the original purpose of this series has now been amply fulfilled, the current article by Jeffrey and Sundaralingam must, perforce, constitute the concluding one of this series. The data recorded are those given in publications made during the years 1979 and 1980, and a complete index is provided for all crystal structures of carbohydrates whose structures were determined from the outset of the technique as applied to carbohydrates (in 1935) through 1980. Furthermore, the article indicates how the computerized data-retrieval systems now provide a ready method for extracting information on structures published subsequent to 1980.

Kensington, Maryland
Columbus, Ohio September, 1985

R. Stuart Tipson
Derek Horton

44. Volume 44

Editors' Preface.—In this volume, M. Mathlouthi (Dijon) and J. L. Koenig (Cleveland) discuss the vibrational spectra of carbohydrates in an article that updates and vastly expands those by W. B. Neely in Volume 12 and by H. Spedding in Volume 19 of this series. Important advances in both infrared and Raman spectroscopy have stemmed from discovery of the fast Fourier-transform algorithm, the introduction of efficient minicomputers, the development of Fourier-transform spectrophotometers, and the use of lasers for Raman spectroscopy.

Although vibrational spectroscopy has been overshadowed for many years by n.m.r. spectroscopy as a tool for studying molecular structure and interactions, the new developments now readily permit normal coordinate analysis of molecules of the complexity presented by carbohydrates, and the technique is of particular importance for studying hydrogen-bonding interactions of carbohydrates.

In an article that collates information not extensively treated before, Z. J. Witczak (West Lafayette) describes the synthesis, chemistry, and preparative applications of monosaccharide thiocyanates and isothiocyanates; the thiocyanate anion is an ambident nucleophile of great synthetic versatility in approaches to nucleoside analogs and to thio and deoxy sugars.

B. V. McCleary (Rydalmere) and N. K. Matheson (Sydney) present a broad discussion of the analysis of polysaccharide structure by use of specific degradative enzymes and bring up to date the treatment of the subject as devoted to D-glucans by J. J. Marshall in Volume 30.

The biosynthesis of bacterial polysaccharide chains composed of repeating units is treated by V. N. Shibaev (Moscow), who coordinates our knowledge of the manner in which nucleoside and polyprenyl glycosyl diphosphates serve to generate polysaccharides of great structural diversity. A complementary discussion, by R. Pont Lezica and G. R. Daleo (Mar del Plata) and P. M. Dey (Egham), treats the role of lipid-linked sugars as intermediates in the biosynthesis of complex carbohydrates in plants. The final article, by N. K. Kochetkov and G. P. Smirnova (Moscow), on glycolipids of marine invertebrates complements that by E. Lederer in Volume 16 on those of acid-fast bacteria, by Y.-T. Li and S.-C. Lion on the biosynthesis and catabolism of glycosphingolipids (Volume 40), and by R. T. Schwarz and R. Datema on the lipid pathway of protein glycosylation and its inhibitors (Volume 40).

Finally, an obituary of Fred Shafizadeh is provided by his former student, G. D. McGinnis.

Kensington, Maryland R. Stuart Tipson
Columbus, Ohio July, 1986 Derek Horton

1924–1983

Fred Shafizadeh

45. Volume 45

Editors' Preface.—This volume pays tribute to two carbohydrate pioneers. H. Stetter highlights the life and work of Burckhardt Helferich, one of the last of Emil Fischer's protégés and a pioneer in protective-group strategy and glycosidic coupling. The scientific career of Francisco García González, presented by his students, A. Gómez-Sanchez and J. Fernandez-Bolaños (Seville) provides a sensitive account of the contributions of Spain's leading carbohydrate chemist whose work emphasized especially the reactions of monosaccharides that generate aromatic heterocycles.

In line with the policy of *Advances* to provide periodic coverage of major developments in physical methodology for the study of carbohydrates, A. Dell (London) here surveys the use of fast-atom-bombardment mass spectrometry in application to carbohydrates. This technique has achieved rapid prominence as the "soft" ionization technique of choice for structural investigation of complex carbohydrate sequences in biological samples. The author's extensive personal involvement in this field makes her chapter a critical, state-of-the-art overview for the specialist, as well as a valuable primer for the reader unfamiliar with this technique.

In contrast to the rapid rise of f.a.b.-mass spectrometry, the use of circular dichroism in the carbohydrate field, here surveyed by W. Curtis Johnson, Jr. (Corvallis, Oregon), has evolved more slowly and been less widely appreciated. Instrumental limitations have hampered the routine use of circular dichroism; most sugars are transparent in the 1000–190-nm wavelength region of c.d. instruments, and the technique has thus invited less use for them than for nucleic acids and proteins, which show strong absorption in this region. However, as Johnson points out, c.d. is a potentially powerful tool for investigation of the stereochemistry of monosaccharides, of intersaccharide linkages, and most importantly, of the secondary structure of polysaccharides. Many valuable applications of this technique evidently lie in the future.

The significance of n.m.r. spectroscopy for structural elucidation of carbohydrates can scarcely be underestimated, and the field has become vast with ramifications of specialized techniques. Although chemical shifts and spin couplings of individual nuclei constitute the primary data for most n.m.r.-spectral analyses, other n.m.r. parameters may provide important additional data. P. Daïs and A. S. Perlin (Montreal) here discuss the measurement of proton spin–lattice relaxation rates. The authors present the basic theory concerning spin–lattice relaxation, explain how reliable data may be determined, and demonstrate how these rates can be correlated with stereospecific dependencies, especially regarding the estimation of interproton distances and the implications of these values in the interpretation of sugar conformations.

Specific applications of carbon-13 n.m.r. spectroscopy to the glycophorins, an important family of glycoproteins present in the human erythrocyte membrane, are discussed by K. Dill (Clemson), who demonstrates the value of ^{13}C-n.m.r. spectra for the structural mapping of glycoproteins.

Finally, C. K. Lee (Singapore) contributes an extensive article surveying the chemistry and biochemistry of the phenomenon of sweet taste. The sweetness of sugars has been of interest since ancient times, but even now a total understanding of this gustatory response remains elusive.

The editors note with regret the passing of Laszlo Mester on February 23, 1986. Mester worked extensively on hydrazine derivatives of sugars, and he contributed a notable article on the formazan reaction in Volume 13 of this series.

Kensington, Maryland R. Stuart Tipson
Columbus, Ohio August, 1987 Derek Horton

1887–1982

Burckhardt Helferich

1902–1983

Francisco García González

46. Volume 46

Editors' Preface.—Traditional chromatographic methods for the separation and purification of carbohydrates of all kinds, ranging from mono- to oligo-saccharides, have permitted many important developments in the carbohydrate field. A major advance that has achieved great utility is high-performance (or so-called high-pressure) liquid chromatography. This technique, treated in comprehensive practical detail in the present volume by Kevin B. Hicks (Philadelphia), affords, within an hour, precise analytical and preparative separations of mixtures hitherto separable only with difficulty.

The next chapter, by Rene Csuk and Brigitte I. Glänzer (Zurich), constitutes an extensive treatise on the nuclear magnetic resonance (n.m.r.) spectroscopy of fluorinated monosaccharides [whose early chemistry was surveyed in Vol. 38 (1981) by Anna A. E. Penglis]; the comprehensive data tabulated herein should be especially of value to those working in the field. It continues the coverage, in *Advances,* of n.m.r. spectroscopy as the key tool for characterization of carbohydrates. It complements articles on the 1H-n.m.r. spectroscopy of carbohydrates by Laurance D. Hall [Vols. 19 (1964) and 29 (1974)], Bruce Coxon [Vol. 27 (1972)], and Johannes F. G. Vliegenthart, Lambertus Dorland, and Herman van Halbeek [Vol. 41 (1983)], and on the ^{13}C-n.m.r. spectroscopy of monosaccharides by Klaus Bock and Christian Pedersen [Vol. 41 (1983)], of oligosaccharides by the same authors and Henrik Pedersen [Vol. 42 (1984)], and of polysaccharides by Philip A.J. Gorin in Vol. 38 (1981).

Protecting groups remain central to the methodology for synthesis of ever more-complex carbohydrate targets. Herein, Uri Zehavi (Rehovot) discusses a somewhat under-utilized but potentially elegant and useful aspect, namely, that of photosensitive protecting groups capable of selective introduction with accessible reagents and subsequent removal under mild irradiation. The chapter is a useful adjunct to that by Roger W. Binkley on the photochemical reactions of carbohydrates that were adumbrated in Vol. 38 (1981).

The next chapter, by Ronald T. Clarke, John H. Coates, and Stephen F. Lincoln (Adelaide) discusses inclusion complexes of the cyclomalto-oligosaccharides (cyclodextrins), a unique group of natural cryptands that has attracted great interest within and outside the carbohydrate field in recent years. The article updates the pioneering contribution by Dexter French in Vol. 12 (1957) on these oligosaccharides, then known as the Schardinger dextrins.

Two chapters treat widely divergent aspects of the aqueous degradation of carbohydrates. Christopher J. Biermann (Corvallis) discusses aqueous acidic hydrolysis and other cleavages of glycosidic linkages in oligo- and poly-saccharides, with specific emphasis on their relation to procedures for determination of chemical structure. In the final chapter, Olof Theander (Uppsala) and David A. Nelson (Richland) provide an informative treatment of the aqueous, high-temperature transformation of starch, cellulose, and other abundant carbohydrates relative to the utilization of biomass as a source of useful chemical feedstocks.

This issue pays tribute to two scientists who pioneered the development of carbohydrate chemistry in their respective countries of Japan and Argentina. Tohru Komano (Kyoto) and Naoki Kashimura (Mie) describe the life and work of Konoshin Onodera, and an obituary article on Venancio Deulofeu is contributed by Rosa M. de Lederkremer and Eduardo G. Gros (Buenos Aires).

Kensington, Maryland
Columbus, Ohio July, 1988

R. Stuart Tipson
Derek Horton

1910–1983

Konoshin Onodera

1902–1984

Venancio Deulofeu

47. Volume 47

Editors' Preface.—The first chapter in Volume 47 is an authoritative account by S. J. Angyal (Kensington, N.S.W., Australia) on complexes of metal cations with sugars in solution. The potential uses of such complexes to control the behavior of sugars are not as well recognized as they might be, and Angyal's account will be of great value for analytical chemists seeking enhanced separation methodology, for spectroscopists studying molecular structure, and for synthetic workers.

A comprehensive survey of the anomeric and exo-anomeric effects in carbohydrate chemistry is contributed by I. Tvaroška and T. Bleha (Bratislava, Czechoslovakia). The stereochemical influence of these effects on the reactions of sugars and on conformational behavior about the glycosidic linkage has far-reaching significance but, as the article points out, our theoretical understanding has not yet been perfected to a fully integrated and quantitatively predictive basis.

The contribution provided by K. Dill and R. D. Carter treats a specialized aspect of metal ion–carbohydrate interaction, devoted to the use of ^{13}C-n.m.r. spectroscopy with Gd^{3+} and Mn^{2+} in particular as shift reagents and relaxation probes to study the behavior of biological carbohydrates that normally interact with Ca^{2+} and Mg^{2+}.

A newer technique that will certainly take its place as a standard tool in polysaccharide structural analysis is the selective depolymerization by anhydrous hydrogen fluoride, here surveyed by Yu. A. Knirel and E. V. Vinogradov (Moscow, U.S.S.R.) and A. J. Mort (Stillwater, OK).

The thermal decomposition of sugars and oligosaccharides to produce caramels for use in foods, and the thermal modification of starch to manufacture "dextrins" for use as adhesives, constitute traditional technological arts of considerable commercial significance, although much of the specific chemistry involved in these processes, as well as in the higher temperature pyrolytic breakdown, remains poorly understood by contemporary standards. In two parts, P. Tomasík, S. Wiejak, and M. Pałański (Poland) bring together a vast amount of information not hitherto readily available to workers in the field. It may be hoped that their efforts will now stimulate further work using modern techniques to furnish precise structural characterization for these materials, which would help in the rational control of manufacturing processes and in the precise definition of the products.

The last chapter, written by S. E. Harding (Nottingham, United Kingdom), describes and discusses the macrostructure of mucus glycoproteins, complex polyelectrolytes whose behavior in solution is governed by aspects of secondary and tertiary structure that control their interactions in biological systems.

Kensington, Maryland
Columbus, Ohio August, 1989

R. Stuart Tipson
Derek Horton

48. Volume 48

Editors' Preface.—Analogues of the cyclic sugars in which the ring-oxygen atom is replaced by a methylene group were studied in 1966 by McCasland, who named them "pseudo-sugars." With the recognition that such compounds have a variety of interesting properties, especially as biochemical probes, and with the development of effective methods for their controlled synthesis and structural characterization, there has been much recent interest in this class of compounds, here surveyed by Suami (Tokyo) and Ogawa (Yokohama), who have themselves contributed a major proportion of the recent literature on these compounds. The "pseudo-sugar" terminology is unfortunately vague and not amenable to indexing. Neither the rational names based on the cyclitol terminology nor the fully systematic Geneva names are readily comprehended in reference to conventional carbohydrate nomenclature. The standard IUPAC "carba" prefix for replacement by carbon of a hetero atom in a compound having a recognized trivial name provides a rational solution to the problem of assigning explicit yet recognizable names to these compounds, and the "carba-sugar" names employed here should provide a superior compromise.

The element fluorine is an atypical halogen, and likewise, the fluorinated sugars are quite different in many respects from other halogenated sugars; their chemical synthesis frequently requires specialized methods. The development of effective new synthetic reagents, coupled with an extraordinary interest in the role of fluorinated sugars in biological processes, has led to an almost explosive growth of activity in this field. Although the subject was treated relatively recently, by Penglis in Volume 38, and the n.m.r. properties of fluorinated sugars were discussed by Czuk and Glänzer in Volume 46, the growth of the field has been so fruitful that the chapter here presented by Tsuchiya (Kawasaki) was obliged to occupy far more space than a normal *Advances* article; it was considered better to retain the subject material integrated into one large article than to fragment it into several shorter chapters.

As early as Volume 2 of this series, the bacterial polysaccharides were surveyed in independent articles by Stacey and by Evans and Hibbert. There was little indication then of the profusion of structural types of sugar compounds and linkage patterns later to be found in these polysaccharides. Earlier analytical methods for separating and characterizing these components were very tedious. In the article presented here, Lindberg (Stockholm) provides a comprehensive yet compact overview of the current state of this greatly expanded field; the identification of close to 100 component sugar structures has largely been made possible through advances in structural analytical methodology pioneered in his own laboratory.

The hydrolysis of glycosides by acid and by enzymes is one of the most important reactions encountered in the carbohydrate field. The mechanisms of acid hydrolysis of glycosides was surveyed by BeMiller in Volume 22, and in this volume, Legler (Köln) provides an authoritative account of the glycoside hydrolases from the viewpoint of their mechanisms of action as probed by studies with various types of substrate analogs that inhibit these enzymes; the 1967 Phillips mechanism for lysozyme action remains of broad validity for the glycosidases in general.

The prominent role played by Japanese investigators in carbohydrate science is underscored by the two substantial chapters by Japanese authors in the current volume. This volume also pays tribute to one of the greatest Japanese carbohydrate scientists, Hamao Umezawa, in the obituary article contributed by Tsuchiya, Maeda, and Horton. Hamao Umezawa dedicated his entire, extraordinarily productive career to the development of antibiotics; his innovative contributions are exemplified by his chapter in Volume 30 of this series on the biochemical mechanism of inactivation of aminoglycoside antibiotics.

With the completion of this volume, one of us, R. Stuart Tipson, terminates his function as Senior Editor and D. Horton continues as Editor. Tipson was a contributor to the founding volume in 1945 and has been a member of the editorial team since Volume 8 in 1954.

Kensington, Maryland R. Stuart Tipson
Columbus, Ohio September 1990 Derek Horton

Hamao Umezawa

49. Volume 49

Editor's Preface.—Tribute is paid here to the contributions in the carbohydrate field of two notable figures, Rezső Bognár and Jean Émile Courtois, in articles, respectively, furnished by A. Lipták, P. Nánási, and F. Sztaricskai (Debrecen), and by F. Percheron (Paris).

Analysis of the tautomeric compositions of reducing sugars in solution by classical polarimetric methods has inherent limitations, but n.m.r.-spectroscopic methods have greatly enhanced our ability to monitor and quantitate such mobile interconversions of sugars. An excellent overview of developments in this field was presented by S. J. Angyal (Kensington, N.S.W., Australia) in Volume 42. However, the rapid progress of new research, with the advent of more sophisticated spectrometers and techniques of data manipulation, has provided the motivation for a supplement, prepared again by Angyal, which updates and complements his earlier chapter and is to be used in conjunction with it.

The synthetic procedures available to the carbohydrate chemist have been largely dominated by standard reactions proceeding by heterolytic processes within a chiral matrix. The preparative utility of radical-mediated reactions has, however, been amply demonstrated in recent years. The chapter contributed here by L. Somsík (Debrecen) and R. J. Ferrier (Wellington), on bromination reactions of carbohydrates proceeding by radical processes integrates the literature related to Ferrier's pioneering work in this area and underscores its excellent potential in synthesis.

Continuing in the synthetic vein, S. David, C. Augé, and C. Gautheron (Paris) present a practical overview of the potential of enzymes as synthetic tools for the general organic chemist. Their chapter, with a well-selected variety of examples, should help the bench-level organic chemist to overcome the classic preconception that enzymes are exclusively the domain of the biochemist working with nanomolar amounts of material. The David–Augé contribution should materially help in opening up the way for enzymes, both free and immobilized, to be used advantageously for preparative access to important and useful sugars and metabolic intermediates.

P. Stoss (Dottikon, Switzerland) and R. Hemmer (Senden, Germany), in their article on the 1,4:3,6-dianhydrohexitols, provide the perspective of the industrial chemist and bring up to date a subject that was treated by Wiggins in Volume 5 of this series and by Soltzberg in the tabular material contributed in Volume 25. These anhydrides are of considerable theoretical interest, but much of the rapidly burgeoning related research is recorded in the patent literature because of the wide practical potential manifested by these bicyclic diols.

Although the classic proteoglycans of cartilage tissue are now well characterized, considerably less is known concerning the "small proteoglycans" containing only one or two glycosaminoglycan chains on the protein core; their structures and biological roles are surveyed here by H. Garg and N. Lyon (Boston).

It is with great regret that I record the passing on July 13, 1991 of R. Stuart Tipson in his 85th year. Dr. Tipson was a contributor to the first volume in this series and a member of the editorial team beginning with Volume 8 in 1954 until his retirement from the editorship at the completion of Volume 48 in 1990. A fuller survey of his life and scientific work is scheduled for an upcoming volume.

Columbus, Ohio August 1991   Derek Horton

Reszõ Bognár

Jean Émile Courtois

50. Volume 50

Editor's Preface.—With this fiftieth volume the *Advances* reaches its half century, during which time the carbohydrate field has evolved dramatically across a broad range of formal scientific disciplines, both fundamental and applied. Over the years the chapters in this series have chronicled these developments for the benefit of the general reader and have concurrently provided, for the specialist, important critical insight into gaps in our knowledge. The rich legacy of the early carbohydrate literature remains a fruitful resource in addressing new problems with today's superior tools of research.

Lemieux and Spohr (Alberta) here trace our understanding of enzyme specificity in broad perspective as they assess Emil Fischer's "lock and key" concept advanced a century ago in relation to current ideas of molecular recognition. It may be noted that the very first article in Volume 1 of *Advances*, by Claude S. Hudson, was devoted to the Fischer cyanohydrin synthesis and the consequences of asymmetric induction.

The task of interpreting chemical transformations and the logical planning of synthetic methods have been traditionally difficult with the carbohydrates because of their polyfunctionality and complex stereochemical architecture. The vast body of empirical literature is daunting to the newcomer to the field, and the synthesis of glycosides by endless permutations of the traditional Koenigs–Knorr synthesis presents especial difficulty. A major step forward has resulted from the insightful thinking of R. R. Schmidt (Konstanz) toward the rational design of practical and versatile methodology for glycoside synthesis. His trichloroacetimidate method, here surveyed in comprehensive detail in a chapter with his colleague W. Kinzy (Basel), constitutes one of the most imaginative approaches to an important problem in synthetic methodology. Their chapter will undoubtedly comprise a key reference source for numerous researchers for many years to come.

Although the use of abundant sugars as starting materials for chiral synthesis has received considerable attention, the ready availability of many aldonolactones is less well recognized by "mainstream" synthetic organic chemists. The chapter here contributed by de Lederkremer and Varela (Buenos Aires) provides a comprehensive overview of the practical potential of these cyclic esters and complements the more specialized contribution on gulonolactones by Crawford in Volume 38.

The biomedical importance of infections by gram-negative pathogens and the consequences of septic shock have drawn much attention to lipid A, the toxic subcomponent of the lipopolysaccharide endotoxin of these organisms. A comprehensive account of the chemical structures and biological behavior of the lipid A structures is presented here by Zähringer, Lindner, and Rietschel. The chapter incorporates much of their own work from the Borstel laboratory where Westphal animated his pioneering work on bacterial lipopolysaccharides.

The sugar–amino acid linkage point in glycoproteins and proteoglycans, involving the side-chain nitrogen atom of L-asparagine or the hydroxyl group of L-serine or L-threonine, is a key structural region of these glycoconjugates. Garg, von dem Bruch, and Kunz (Boston and Mainz) now survey developments in the synthesis of glycopeptides containing these linkages, updating with significant new work the earlier reports in Volume 25 by Marshall and Neuberger and in Volume 43 by Jeanloz and Garg.

The final chapter, by Hounsell (London), also relates to an important aspect of glycoprotein structure, namely the structures and shapes, as determined by physicochemical methods, of oligosaccharide determinants of glycoproteins that are antigens and targets for binding of adhesion molecules.

For most of its existence the *Advances* has been guided by the breadth of scientific insight and editorial expertise of two individuals, M. L. Wolfrom and R. S. Tipson. The obituary chapter in this volume records the life and scientific work of Robert Stuart Tipson, who contributed a chapter on the nucleic acids to Volume 1 and retired as Editor with the publication of Volume 48.

Washington, D.C., April 1994   Derek Horton

Robert Stuart Tipson

51. Volume 51

Editor's Preface.—Structural sugar chemistry received its key foundations in work by German, British, and American investigators during the late 19th and the first half of the 20th century. These investigators employed the classical tools of organic chemistry in conjunction with the physical technique of polarimetry and pioneering applications of X-ray crystallography. However, many complex problems of tautomerism posed by sugars and their derivatives, along with questions concerning their conformations and reactivities, resisted solution until the advent of more powerful analytical methods. It is noteworthy that development of the most significant of these techniques, NMR spectroscopy, as it rose to its prominent role as a structural tool in chemistry and biochemistry, owes much to the carbohydrate field and to the seminal work in the 1950s of R. U. Lemieux.

Considered a mature technique in all areas of science, NMR spectroscopy has passed from being used for the simple recording and interpretation of chemical shifts and spin couplings for small molecules to the domain of specialists performing a multitude of procedures designed to extract detailed NMR parameters from biomolecules of ever-increasing complexity. In this volume of *Advances,* two chapters are devoted to current aspects of NMR spectroscopy. Tvaroška and Taravel (Bratislava and Grenoble) discuss carbon–proton coupling constants from both the experimental and theoretical points of view and address the significance of these couplings in elucidating the conformational behavior of sugars and their derivatives. The extensive tabular information relating these couplings to dihedral angles in conformationally rigid molecules will constitute an important source of reference for other workers probing the three-dimensional structures of carbohydrates in solution.

A complementary article by Daïs (Iraklion, Crete) addresses the theoretical principles underlying the phenomenon of carbon-13 nuclear magnetic relaxation, encompassing spin–lattice (T_1) and spin–spin (T_2) relaxation times, the nuclear Overhauser enhancement, and their relation to the motional behavior of carbohydrates in solution. With examples broadly selected from simple sugar derivatives, oligosaccharides, and polysaccharides, the author shows how qualitative treatments have provided useful interpretations of the gross mobility of molecules in solution, but demonstrates how a quantitative approach may be of greater ultimate value.

Advances has regularly featured articles focusing on specific enzymes or enzyme groups acting on carbohydrates. The article by Robyt (Ames, Iowa) in this volume is devoted to the glucansucrase enzymes and the mechanism whereby they utilize sucrose to transfer glucose to acceptors and elaborate polysaccharides of the dextran class, including the commercially important dextran from the B-512F strain of *Leuconostoc mesenteroides*.

Aspinall (York, Ontario) and Chatterjee and Brennan (Fort Collins, Colorado), in their article on the surface glycolipids of mycobacteria, focus on the chemistry and biology relating especially to an old medical problem, tuberculosis. This disease remains the leading cause of death from a single infectious agent. After years of decline in industrialized countries, TB is now showing a rapid increase that has some correlation with the incidence of HIV infection. The lipo-oligosaccharide antigens of these bacteria are of extraordinary complexity and variety posing, until recently, major problems of separation and structural characterization. This article brings together a comprehensive survey of isolation methodology and structural detail with a wealth of synthetic virtuosity that has been applied to the construction of oligosaccharide haptens of these antigens. Elaboration of such synthetic oligosaccharide derivatives into neoglycoconjugates has afforded a major tool for probing antibody–antigen interactions and for developing immunological procedures in general.

The final chapter by Tomasík and Zaranyika (Harare, Zimbabwe) deals with an aspect of carbohydrate technology that complements the article by Tomasík and co-workers in Volume 47 on the thermal decomposition of starch. The survey presented in this volume deals with the modifications taking place in starch when it is irradiated by various energy sources, ranging from ionizing radiation to ultrasound, and by such treatments as freeze–thaw cycles and dehydration. Such treatments may impart little visible change to native starch granules, but may profoundly affect the behavior of the material in important technological and nutritional applications.

The life and work of Horace S. Isbell, who ranked with Claude S. Hudson and Melville L. Wolfrom as one of the great American carbohydrate chemistry pioneers, are presented here by El Khadem (Washington, D.C.). During his long career, Isbell made important fundamental discoveries in conformational analysis and in neighboring-group reactions, advances that were not at the time properly recognized by "mainstream" chemists. These were but a part of a sustained dedication to understanding the mechanisms of reaction of sugars and their derivatives, especially through the use of isotopically labeled compounds.

The editor notes with regret the passing of Maurice Stacey, one of the great names of the British carbohydrate school, successor to W. N. Haworth at the University of Birmingham, and one long associated with *Advances in Carbohydrate Chemistry*. A detailed account of his life and work is reserved for a forthcoming volume in this serial.

Washington, D.C. July 1995   Derek Horton

Horace Smith Isbell

52. Volume 52

Editor's Preface.—In his preface to Volume 8 of *Advances*, published in 1953, M. L. Wolfrom, the founding editor, noted that "Carbohydrate nomenclature has been an ever-present problem in this series . . ." and drew attention to the agreement between American and British carbohydrate chemists that resulted in the published "British–American Rules of Carbohydrate Nomenclature." A revision of that document was published in 1962, to be followed seven years later by an internationally proposed set of guidelines for naming carbohydrates and their derivatives.

Since the early 1970s a panel convened by the International Union of Pure and Applied Chemistry and the International Union of Biochemistry and Molecular Biology has been working to formulate recommendations for carbohydrate nomenclature that meet developing needs of research and electronic data handling, while retaining links to the established literature base on carbohydrates.

The realization of these endeavors is presented here in the final document "Nomenclature of Carbohydrates," which provides a definitive reference for current researchers, both in the text version and in the version accessible on the World Wide Web (http://www.chem.qmw.ac.uk/iupac/2carb), where amendments and revisions are maintained.

Garegg (Stockholm), in his chapter on thioglycosides as glycosyl donors, presents a wealth of practical detail on a technique of wide utility for constructing complex oligosaccharides. Much of the work is from his own laboratory. His article complements that by Schmidt in Volume 50 on the trichloroacetimidate method of glycoside synthesis. These two articles chronicle important advances that have been made in the chemical construction of larger oligosaccharides.

Glycosidic coupling methodology nevertheless still falls far short of synthetic methods now standard for oligopeptides and oligonucleotides, where automated syntheses based on solid-phase procedures are routine. There remains considerable scope for further development.

Manley-Harris and Richards (Missoula, Montana) have compiled a comprehensive account of the dianhydrides of D-fructose and related compounds, more than 30 in all. These compounds, several of which are of importance in the sugar industry, have in the past presented significant problems in their chemical characterization. Their chemistry was surveyed as early as 1945 by McDonald in Volume 2 of this series, and discussed again in Volume 22 by Verstraeten. The current article furnishes detailed NMR data for each of the anhydrides, providing definitive reference data for accurate identification and correlation with earlier literature, where erroneous structural attributions are rather frequent.

The vitamin thiamine may not at first sight have a close relation to carbohydrates, but David and Estramareix (Paris) trace here a remarkable story in the elucidation of its biosynthesis. Quite different pathways are shown to exist in prokaryotic and eukaryotic organisms, each involving sugar intermediates, and comparisons offer interesting insight into pathways of biochemical evolution.

X-Ray diffraction analysis of oriented polysaccharide fibers has had a long history. Marchessault and Sarko discussed this topic in Volume 22 of *Advances*, and a series of articles by Sundararajan and Marchessault in Volumes 33, 35, 36, and 40 surveyed ongoing developments. The comprehensive account presented here by Chandrasekaran (West Lafayette, Indiana) deals with some 50 polysaccharides, constituting a wide range of structural types, where accurate data and reliable interpretations are available. The regular helical structures of the polysaccharide chains, and associated cations and ordered water molecules, are presented in each instance as stereo drawings and discussed in relation to observed functional properties of the polymers.

The final chapter, by Clarke, Edye, and Eggleston (New Orleans, Louisiana), deals with the centuries-old technological problem of maximizing yield in the extraction of sucrose from cane or beet juice. Somewhat remarkably, important misconceptions about the fundamental aspects of alkaline degradation of sucrose still persist. The authors of this chapter effectively interpret traditional sugar technology, based largely on empirical art, in clear terms of accepted fundamental principles of chemistry.

The most influential and accomplished British carbohydrate chemist since Sir Norman Haworth was his protégé, Maurice Stacey, who in his long career made broad contributions that bridged chemistry and biology long before the interdisciplinary approach came into vogue. The story of Stacey's life and work, detailed here by Finch and Overend (London), paints a warm picture of a man whose contribution in motivating many young scientists into research careers on carbohydrates was as significant as his own wide-ranging research program.

With regret, the passing is noted, on April 16, 1997, of Guy G. S. Dutton, a member of the Board of Advisors of *Advances* for many years and a staunch supporter of the series. Also deceased in 1997 are two giants of science, Melvin Calvin and Alexander Todd, Nobel laureates, each of whom made seminal contributions in the carbohydrate field.

Washington, D.C.
April 1997

Derek Horton

Maurice Stacey

53. Volume 53

Editor's Preface.—Synthetic work in the carbohydrate field has greatly benefited from new developments in protecting-group strategy and from the introduction of reagents based on heavier elements in the periodic table. Sugars are ideal models for the evaluation of reagents that can specifically bridge two or more oxygen sites. Developments during the past two decades have demonstrated the great value of organotin compounds, especially trialkylstannyl ethers and dialkylstannylene acetals, for effecting useful, high-yielding transformations with high regioselectivity. This volume of *Advances* features a comprehensive survey, by Grindley (Halifax, Nova Scotia), on the application of tin-containing intermediates to carbohydrate chemistry.

In a comparable vein, the element selenium has found important applications in effecting organic transformations. The chapter by Witczak (Storrs, Connecticut) and Czernecki (Paris) on the synthetic applications of selenium-containing sugars surveys key aspects of this still rapidly evolving area. Much of the older work largely parallels the chemistry of the sulfur analogues, but newer developments show novel and versatile potential, particularly in the use of arylselenium reagents for stereochemically controlled addition to glycals and in transformations at the anomeric center.

Molecular recognition between carbohydrates and proteins has become a vast area of broad significance in biochemistry and molecular biology. Anti-carbohydrate antibodies have importance in both medicine and technology. The preparation of protein antibodies having specificity for carbohydrate antigens requires application of a range of specialized techniques in molecular separation, the conjugation of carbohydrate ligands to suitable carriers for immunizing animals, and the isolation of pure antibodies from hyperimmune sera. In this volume, Pazur (University Park, Pennsylvania), a seasoned *Advances* author, contributes an article on anti-carbohydrate antibodies with specificity for monosaccharide and oligosaccharide units of antigens. He provides a practical focus on experimental methodology in preparing antibodies having specificity for various sugars and polysaccharides, with many examples developed in his own laboratory.

It has been almost two centuries since the first reports on the interaction between iodine and starch to produce a blue color, and more than fifty years since the discovery of the complexation between aliphatic alcohols and starch that revealed this polysaccharide to have a linear and a branched component. A large and diffuse body of published work exists on the interaction of starch and its components with other molecules and with ionic substances. Two closely related chapters in this volume, by Tomasík (Cracow, Poland) and Schilling (Ames, Iowa), provide comprehensive accounts of the extensive literature on complexes of starch with inorganic and with organic guests.

Researchers working in industrial or government laboratories often have to rely on their individual experimental efforts in the laboratory and have frequently received less recognition for their accomplishments than have leading scientists in academia, whose work is usually based on extensive experimental work by teams of research students. The three individuals whose lives and scientific contributions are recalled in this volume all made notable discoveries in the carbohydrate field while working in government laboratories and conducting their own experimental work.

John E. Hodge, whose career is reviewed by Feather (Columbia, Missouri), laid the major foundations of the chemical basis of the nonenzymatic browning (Maillard) reaction. In the same Peoria laboratory of the United States Department of Agriculture, Allene Jeanes, whose biography is presented by Roscher (Washington, D.C.) and Sandford (Santa Monica, California), conducted pioneering work on microbial polysaccharides that laid the basis for the commercial development of dextran and xanthan. The memoir contributed by El Khadem (Washington, D.C.) details the career of Harriet L. Frush, whose teamwork with her mentor, Horace S. Isbell, at the US. National Bureau of Standards and later at The American University in Washington, D.C., spanned over sixty years and included seminal contributions in concepts of reaction mechanisms and on the synthesis of labeled sugars.

With this volume we welcome David R. Bundle to the Board of Advisors and look forward to valuable input from the prestigious carbohydrate laboratory in Edmonton, Alberta. With regret we record the death, on April 16, 1997, of Elizabeth Percival, a noted authority on marine algal polysaccharides, and on October 14, 1997, of Iqbal R. Siddiqui, who authored the article "Sugars of Honey" in Volume 25 of this series.

Washington, D.C.   Derek Horton
February 1998

John E. Hodge

Allene Jeanes

Harriet L. Frush

54. Volume 54

Editor's Preface.—This 54th volume of *Advances in Carbohydrate Chemistry and Biochemistry* is an index volume that covers all previous volumes in the series, with a cumulative subject index of the content of all published chapters, along with an index of authors contributing the chapters.

The series began with the 1945 appearance of Volume 1 of *Advances in Carbohydrate Chemistry* under the editorship of Ward Pigman and Melville Wolfrom, and was the first serial publication of a new company named Academic Press, founded by Kurt Jacoby. The stated policy was to have "the individual contributors furnish critical, integrating reviews rather than mere literature surveys, and to have the articles presented in such a form as to be intelligible to the average chemist rather than only to the specialist." The series was to cover the broad field of carbohydrates, including sugars, polysaccharides, and glycosides, and to include biochemical, industrial, and analytical developments.

The founding policy has been sustained throughout the subsequent evolution of the series, under the editorship of Wolfrom for most of the volumes until his death in 1969, of R. Stuart Tipson from 1954 until 1990, and of the present editor since 1969. The essential component of biochemical aspects was emphasized when the title was changed, in 1969, to *Advances in Carbohydrate Chemistry and Biochemistry*. The field of carbohydrates has undergone enormous expansion during the half-century that *Advances* has been in existence, and the pages of the past 53 volumes have recorded important developments in practically all aspects of carbohydrate science. Some of the articles have constituted status reports on areas still under active expansion, while others record definitive information of permanent reference value.

To enhance the utility of *Advances* as a reference source, the provision of good indexes has been an important objective, and each volume has contained a comprehensive name index of all authors cited in the individual chapters, along with a detailed subject index. Responsibility for compiling the subject indexes from the outset and for more than four decades was largely entrusted to a leading expert on indexing and nomenclature, Dr. Leonard T. Capell of Chemical Abstracts Service. These subject indexes were integrated into cumulative indexes of principal topics and have appeared in selected issues, with volume 29 containing such an index for volumes 1–29, while those appearing in volumes 35, 40, 45, and 50 refer in each case to the preceding five volumes.

The present index volume largely follows the model used successfully for the *Methods in Enzymology* series published by Academic Press, and the cumulative subject index has been compiled directly from subject indexes in the individual volumes. It is recognized that a lack of total uniformity is inevitable as a consequence of unevenness in some of the indexes, and most importantly, from the progress that has been made toward uniformity in carbohydrate nomenclature. Current nomenclature recommendations are recorded in the 1996 document "*Nomenclature of Carbohydrates*," issued by the International Union of Pure and Applied Chemistry and the International Union of Biochemistry and Molecular Biology, which has been published in Volume 52 of *Advances in Carbohydrate Chemistry and Biochemistry*.

This index volume is complemented by a cumulative contributor index, and it is hoped that the volume will facilitate the retrieval of information and significantly enhance the utility of the entire series.

Washington, D.C.
June 1999

Derek Horton

55. Volume 55

Editor's Preface.—This issue of *Advances* features an extensive article by García Fernández and Ortiz Mellet (Seville, Spain) on *N*-thiocarbonyl derivatives of carbohydrates. The classic role of these highly reactive derivatives in the synthesis of heterocyclic structures attached to sugar residues has in recent years expanded greatly and now touches on many areas of glycobiology. These derivatives provide simple and efficient methods for tailored construction of complex structural targets, as already well exemplified in established procedures for preparation of neoglycoproteins; newer and developing aspects include potential applications in such areas as solid-phase synthesis and combinatorial chemistry.

A chapter contributed by Varela and Orgueira (Buenos Aires, Argentina) deals with synthetic polyamides formed from sugar derivatives containing amino and carboxyl functionalities. Such chiral analogues of nylon present interesting structural aspects stemming from stereochemical differences in the monomers, and at the practical level they provide a mode for conferring hydrophilicity, biocompatibility, and biodegradability on the polymers.

The classic work of Emil Fischer more than a century ago introduced the chemistry of sugars reacting with phenylhydrazine as a key tool in elucidating stereochemical relationships, and subsequent years revealed a rich variety of products arising from the reactions of different hydrazines with sugars. The structural identity of many of these compounds remained controversial for many years, however, until the advent of modern spectroscopic and X-ray techniques. In a comprehensive article that reflects a career-long preoccupation with this subject, El Khadem, in collaboration with Fatiadi (Washington, D.C.), surveys the entire modern literature on the multifarious compounds resulting from reactions of sugars and their carbocyclic analogues with hydrazine derivatives, emphasizing important applications in synthesis and pointing out areas ripe for further exploration.

One of the oldest areas of enzymology is the action of glycosylases on carbohydrate substrates, but precise mechanistic and stereochemical interpretation of the behavior of these enzymes is still lacking. With the aid in particular of extensive recent X-ray structural work on glycoside hydrolases and studies with substrate analogues, Hehre (New York) here presents detailed insight into the stereochemical factors at work in the action of the "inverting" and "retaining" enzymes.

Volume 52 of this series featured the comprehensive IUPAC–IUBMB document "*Nomenclature of Carbohydrates*" that meets a long-standing need for up-to-date standardized nomenclature for sugar derivatives and complex saccharides. Ongoing work of the international committee has now brought about the updated set of recommendations "*Nomenclature of Glycolipids,*" which is published in this volume.

The life and work of two Nobel Prize winners who made major contributions to the carbohydrate field are recorded here by Buchanan (Bath, UK) writing on Alexander Lord Todd and by Moses (London) with an article on Melvin Calvin.

Godshall (New Orleans) provides an appreciation of Margaret Clarke, whose global influence in the field of sugar technology was but one aspect of her remarkable personal talents in bringing together carbohydrate scientists around the world in constructive synergy.

Washington, D.C. Derek Horton
September 1999

Alexander R. (Lord) Todd

Melvin Calvin

Margaret A. Clarke

56. Volume 56

Editor's Preface.—This fifty-sixth volume of *Advances* features two articles devoted to synthetic aspects. Vaino (La Jolla, CA) and Szarek (Kingston, Ontario) survey the versatility of the heaviest accessible elemental halogen, iodine, for effecting a wide variety of useful transformations in the carbohydrate field under particularly mild conditions.

Reactions involving free radicals, long underutilized in the carbohydrate field, have received considerable attention in recent years, notably through the work of the late D. H. R. Barton and his coworkers, and by the group of B. Giese. Here Praly (Lyon, France) contributes an extensive survey of radicals at the anomeric center of sugars, from both the structural viewpoint and with respect to their chemical transformations, especially under reductive conditions.

Although broad-spectrum antibiotics have long held center stage in the clinical treatment of bacterial infections, the use of carbohydrate vaccines for immunizing subjects against specific pathogens dates back still further, and is once more coming to the forefront. High current interest is focused on the use of specific oligosaccharide determinants chemically conjugated to proteins as vaccine candidates, an area here discussed by Pozsgay (Bethesda, MD).

The complex structures of mammalian proteoglycans continues to challenge researchers using even the most sophisticated current methodology. The glycosaminoglycan and protein subunits in these glycoconjugates are connected through a specific oligosaccharide linkage-region, and the molecular structure of this region is the subject of the article by Rama Krishna and Agrawal (Birmingham, Alabama).

The final chapter in this volume, by Jiménez-Barbero, Espinosa, Asensio, Cañada, and Poveda (Madrid, Spain), is devoted to the conformations of *C*-glycosyl compounds, and focuses on the complementary use of computational methodology and experimental studies, largely by NMR, to probe the conformational properties of this class of glycoside analogues, which are of high interest in their behavior with glycoside hydrolases and transferases.

The passing is noted with regret of three major figures in the carbohydrate world, George A. Jeffrey on February 13, 2000, Sumio Umezawa on March 30, 2000, and Raymond U. Lemieux on July 22, 2000. All three were authors of important articles in this Series, and detailed obituaries are scheduled to appear in a future volume.

An appreciation of the life and work of Guy G. S. Dutton, a frequent contributor to this Series and a member of its Advisory Board, is contributed by Parolis (Constantia, South Africa).

Washington, D.C. Derek Horton
May, 2000

Guy G.S. Dutton

57. Volume 57

Editor's Preface.—The 57th volume of Advances reflects sustained trends in the carbohydrate field toward strong emphasis on biological aspects, both fundamental and applied, and a major proportion of the current issue is devoted to this viewpoint. The article by Monteiro (Ottawa, Ontario) on the lipopolysaccharides of *Helicobacter pylori* presents a comprehensive account of the chemical structure, biosynthesis, and potential pathogenic role of these bacterial cell-surface components in a wide range of disorders of the human stomach. It is noteworthy that less than two decades have elapsed since this microbiological basis for gastric ulcers and related diseases became recognized. The structural work presented here reveals in particular a remarkable similarity between the saccharide O-chains of the lipopolysaccharides and the sialyl-Lewis human blood-group antigens, a factor that may aid in the survival of the organism in the host and which presents a major challenge in the development of effective therapeutic vaccines.

Fifteen years ago, Benito Casu (Milan) contributed an authoritative article to this series on the structure and biological activity of heparin. Although this glycosaminoglycan has been used clinically for many decades in treating thromboembolic diseases, precise details of its mode of action have been lacking. Major developments in analytical methods and in our understanding of biosynthetic processes, together with recognition of the complementary role of heparan sulfate in polysaccharide–protein binding interactions involving antithrombin and other proteins, have permitted understanding of these interactions in much greater molecular detail. These developments, and their significance in our understanding of cell-surface interactions and in therapeutic applications, are presented here by Casu and Lindahl (Uppsala).

Unger (Vienna) contributes a major article devoted to physiological aspects of carbohydrates, in a wide-ranging survey of the oligosaccharide ligands of mammalian lectins (selectins) that function as cell-adhesion receptors. The recognition of the essential role of selectin–carbohydrate interaction in the inflammatory response and the key involvement of the sialyl-Lewis^x oligosaccharide and its analogues in the carbohydrate determinant has stimulated an enormous research effort on many fronts in the quest for effective therapeutic agents. This has led to great advances in the three-dimensional understanding of carbohydrate structures and their protein-binding domains, although the weakness of these interactions still presents difficulties in the design of practical anti-inflammatory drugs.

Replacement of the ring oxygen atom in sugars by other atoms, especially by sulfur or nitrogen, has long presented a challenge to synthetic chemists, and the recognition that such analogues have potential as inhibitors of glycoprocessing enzymes has stimulated much recent activity in this area. In this volume, Fernandez-Bolaños and Maya (Seville), together with Al-Masoudi (Konstanz), survey the chemistry of sugars having sulfur in the ring, leading to products that have shown useful potential as, for instance, oral antithrombotic agents, antidiabetic agents, and agents for treatment of HIV infections.

The lives and scientific work of two legendary figures in the carbohydrate field are commemorated in this issue. George A. Jeffrey was a pioneer crystallographer whose doctoral work in the laboratory of Haworth on the structure of glucosamine hydrobromide, a monumental task at the time, firmly established the chair conformation of the pyranose sugars and set the stage for the stereochemical correlation of the carbohydrates with the amino acids and proteins. The account presented here by French (New Orleans) documents his work during six decades in developing and applying crystallographic techniques for characterization of sugars and their derivatives in the solid state. He was the undoubted authority on all aspects of the molecular geometry of crystalline sugars and for many years contributed extensively to these *Advances*.

The article by Miyake (Kawasaki), Maeda (Tokyo), and this writer details the long career of Sumio Umezawa devoted to the chemistry and medicinal applications of antibiotics, especially the aminocyclitols (aminoglycosides), a field dominated by Sumio and his microbiologist/biochemist brother Hamao from the earliest days of streptomycin through to practical semisynthetic analogues developed by Sumio that have enjoyed wide clinical application. The complementary articles by the two Umezawa brothers in Volume 30 of this series remain a definitive reference work on these antibiotics.

With this volume we welcome David C. Baker to the Board of Advisors.

Washington, D.C. Derek Horton
July, 2001

George A. Jeffrey

Sumio Umezawa

58. Volume 58

Editor's Preface.—This 58th volume of *Advances* continues the sustained theme of definitive articles written by leaders in the field, providing comprehensive coverage of mature fields or selective treatment of evolving areas. In contrast to the strong emphasis on glycoconjugates and biological aspects in the preceding Volume 57, the contents of the present issue reflect mainly structural and synthetic aspects of simpler sugars.

Methodology for glycosidic coupling continues to pose challenges for the synthetic chemist, and the synthesis of defined oligosaccharide sequences remains a problem requiring great skill and experimental versatility. In contrast to the effective and widely available automated procedures based on solid-support technology for synthesis of oligopeptides and oligonucleotides, comparable procedures in the carbohydrate field have been lacking, on account of the complexities of linkage position, anomeric orientation, and protecting-group manipulation. Among the investigators addressing this problem, Seeberger (Cambridge, Massachusetts) and his coworkers Plante and Palmacci have made significant strides in developing the functional solid-phase automated synthesizer presented here. It provides practical feasibility for synthesis of selected oligosaccharide sequences, although further research to permit general application for other linkage patterns clearly remains necessary.

Although sugar derivatives containing unsaturated functionality were introduced back in Emil Fischer's time, and the rather inappropriate name "glycal" became a term that persists to this day, much of their chemistry remained confusing until the advent of NMR spectroscopy. Early surveys in Volumes 7 and 9 of this series, by Freudenberg and Blair, respectively, were followed notably by Ferrier's landmark article in Volume 20, which demonstrated the exceptional versatility of unsaturated sugars in synthesis. Ferrier augmented his article soon afterwards, in Volume 24. The innovative leadership of Ferrier in this area of synthesis has become legendary, and here he and his colleague Hoberg (Lower Hutt, New Zealand) revisit the subject of unsaturated sugars in a definitive treatment from the current viewpoint.

A comprehensive survey of all classes of internal anhydrides of sugars is provided in this volume by Černý (Prague), encompassing cyclic sugars bridged by three-, four-, and five-membered oxygenated rings. Many such derivatives offer important potential in synthesis. Earlier articles in this series, especially those by Peat in Volume 2 and by Černý and Staněk in Volume 34, still provide useful background and older detail on these derivatives, as does the 1972 survey by Guthrie in Volume 1A of "The

Carbohydrates, Chemistry and Biochemistry," edited by Pigman and Horton (Academic Press).

Two related chapters, contributed by de Lederkremer and Marino, and by Varela, both from Buenos Aires, deal with the processes and products of oxidation of carbohydrates, and offer extensive updating of the 1980 article by Green on acids and other oxidation products of sugars, and the one by Theander on oxidative and degradative reactions of sugars and polysaccharides, both published in Volume 1B of the Pigman–Horton treatise. These two articles from Argentina offer broad coverage of all aspects of carbohydrate oxidation, from both the fundamental view and from technological considerations. As an aid to the reader, titles of the cited articles are included in the extensive bibliographic references. Titles are also incorporated in the references cited in the Ferrier–Hoberg chapter. It is proposed to incorporate such titles on a standard basis in future volumes in this series.

The great biological significance of the sialic acids, nine-carbon 5-amino sugars, has long been recognized, and Schauer provided a definitive survey of their chemistry and biochemistry in Volume 40. Much more recently, the wide occurrence in microorganisms of related nonulosonic acids aminated also at position 7 has been demonstrated. Major work on these diamino sugars by three groups has led to the collaborative chapter by Knirel, Shashkov, and Tvetskov (Moscow), Jansson (Huddinge, Sweden), and Zähringer (Borstel, Germany) featured here as the last contribution to this volume.

The life and work of one of the greatest carbohydrate scientists of our time, Raymond U. Lemieux, is recalled here in a sensitive account by Bundle (Edmonton). During a remarkably productive career extending over more than half a century, Lemieux pioneered the application of NMR spectroscopy in chemistry, developed rational approaches for glycosidic coupling, made major contributions to our understanding of three-dimensional carbohydrate structures and protein binding, and made important contributions in the biomedical area. His own articles in these *Advances* include the chemistry of streptomycin in Volume 3, the mechanisms of replacement reactions in Volume 9, and in Volume 50 a consideration of Emil Fischer's "lock and key" concept of enzyme specificity.

With this present volume we welcome Peter H. Seeberger and Yuriy Knirel to the Board of Advisors.

Washington, D.C. Derek Horton
August 2003

Raymond Urgel Lemieux

59. Volume 59

Editor's Preface.—Many recent volumes in this series have been strongly oriented toward biochemical or medicinal chemical aspects. This current volume focuses on basic chemical aspects of mono- and oligo-saccharides, and on polysaccharide technology. de Lederkremer and Gallo-Rodriguez (Buenos Aires) provide a comprehensive survey of those monosaccharides (excluding amino and deoxy sugars) that are found in natural products, in a treatment that greatly extends a 1972 treatment of this subject by Schaffer.

The chapter by Garegg (Stockholm) presents a broad overview of the chemistry of glycosides, dealing with their synthesis, structure, and properties. It places particular emphasis on the ongoing challenge of glycoside synthesis as it applies to complex oligosaccharide targets, and it features much of the author's own considerable contribution to this area.

The inositols and their derivatives can be regarded as monosaccharide sugars in which the ring oxygen atom has been replaced by carbon, and they may thus be named as carba sugars. The inososes derived by their oxidation react with hydrazines in transformations that are often of considerable complexity, thus resembling the behavior of sugars. The hydrazine derivatives of sugars have been surveyed in detail in this Series by El Khadem in Volume 20 and again in conjunction with Fatiadi in Volume 55. In this volume El Khadem and Fatiadi (Washington, D.C.) complement this aspect with a comprehensive account of the hydrazine derivatives of carba sugars.

The technology of polymeric carbohydrates is strongly oriented to the most abundant examples, namely starch and cellulose. Tomasík (Cracow) and Schilling (University Center, Michigan), in their wide-ranging article on chemical derivatization of starch, present an extensive compilation of the literature on potentially useful products formed by esterification, etherification, oxidation, and other reactions with starch. Much of the literature cited comes from patent sources, not subject to the conventional refereeing procedures in effect for journal articles, and so the reader needs to judge appropriately the validity of some of the claims made for product structure and practical application.

The life and work of Edward J. Hehre, a pioneer in our knowledge of the mechanism of action of glycosylase enzymes, is the subject of the article by Brewer (New York).

It is noted with regret the passing on February 23, 2004, of Aleskander Zamojski, who contributed in Volume 40 an article on the synthesis of sugars from noncarbohydrate sources, and on July 31, 2004, of Jacques van Boom (Leiden), who made important contributions to carbohydrate synthesis and to glycobiology.

With this volume we thank two members of the Board of Advisors, Roy L. Whistler and Bengt Lindberg, for their long service to the series, and welcome Geert-Jan Boons and Serge Pérez to the Board.

Washington, D.C.   Derek Horton
September 2004

Edward J. Hehre

60. Volume 60

Editor's Preface.—Amino sugars continue to play a major role in the overall carbohydrate field, and they present particular challenges for the synthetic chemist seeking to effect useful and controllable transformations toward targets of biological importance. Presented in this volume are two articles that provide complementary viewpoints on methodology for the manipulation of nitrogen functionality in sugar derivatives. The article by Karban and Kroutil (Prague) offers for the first time a comprehensive account of the chemistry of sugar aziridines (epimines), emphasizing preparative methods for introduction of the three-membered aziridine ring into a sugar framework, and on ring-opening reactions under controlled conditions to afford defined targets.

Glycosyl azides form the focus of the article by Györgydeák (Debrecen) and Thiem (Hamburg), furnishing a detailed survey of methods for introducing the azide group at the anomeric center, together with the wide range of transformations possible with this versatile and highly reactive functional group in both the monosaccharide framework and also in complex oligosaccharide structures related to glycopeptides and glycoproteins. Sadly, Zoltán Györgydeák, a prolific contributor to the carbohydrate literature and coauthor of the book "*Monosaccharide Sugars*" (1998, Academic Press) died after this article was completed.

When in 1928 Louis Malaprade described the oxidation of ethylene glycol and some other polyalcohols with periodic acid, he could never have envisaged the importance that this glycol-cleavage reaction was later to attain in the carbohydrate field. As a tool for determining structure in polysaccharides and glycoconjugates it now features in so many publications that it would be impossible to cover all of its applications in a single article. The fundamentals of the glycol-cleavage reaction, with both periodate and lead tetraacetate, in all types of carbohydrate structure, are surveyed authoritatively here by Perlin (Montreal), himself a pioneer who has made major contributions to our knowledge of the subject. His article provides a thorough basis for understanding the scope and potential limitations in applying the reaction, and gives a clear explanation of the widely used (but frequently misunderstood) Smith degradation procedure for structure determination.

The amino sugar theme reappears in a very different context in the article by Willis and Arya (Clemson, SC) dealing with the aminoglycoside antibiotics, some sixty years after the discovery of streptomycin and three decades after the landmark articles in Volume 30 of this series by the Umezawa brothers who contributed so much to our knowledge of structure, synthesis, and mechanism of bacterial resistance to these antibiotics. The authors here explore the complex area of interaction of aminoglycosides with nucleic acids, from early concepts of ribosomal binding and disruption of the translation message to mRNA to our present recognition of the multiplicity of the interactions of these basic molecules with many forms of RNA and also DNA. This understanding has key significance in efforts to develop new, less toxic, therapeutic agents effective against resistant bacteria.

Many problems in current carbohydrate science require a multidisciplinary approach, as demonstrated here by the joint Spanish–German group coordinated by Jiménez-Barbero (Madrid) and a large group of coauthors, who address the question of noncovalent interactions between carbohydrates and proteins. In the article they explore both hydrophilic and hydrophobic interactions, as revealed by a wide range of experimental and computational methods, with particular emphasis on the defense proteins (lectins) of plants and their interactions with the chitin of fungi and other plant pathogens. They show that the lectins share a common structural motif (chitin-binding domain or hevein domain) involved in the defense mechanism, and provide a particularly useful model for studies at the atomic resolution level of the hydrogen-bonding and carbohydrate–aromatic interactions between proteins and carbohydrates.

The biographical article by Lundt and Bock (Copenhagen) gives an account of the life and work of Christian Pedersen, whose contributions on the use of anhydrous hydrogen fluoride as a solvent for studying the reactions of carbohydrates are particularly notable. In an era when most leaders in the field direct the work of large groups of coworkers, Pedersen harks back perhaps to the days of Emil Fischer in that he conducted much of his work with his own hands, and introduced many preparatively useful synthetic procedures without recourse to chromatography.

Much of the early scientific work of the late Aleksander Zamojski focused on the total synthesis of racemic monosaccharides, based on stereocontrolled reactions of substituted dihydropyrans obtained by Diels–Alder cycloaddition. His article in Volume 40 of this series details many of his early studies in this area. His colleagues Jarosz and Chmielewski (Warsaw) here offer a broad insight into Zamojski's contributions, which later extended into chiral structures, higher sugars, and oligosaccharides.

The passing in 2005 is noted of Nikolai Kochetkov, a major figure on the world carbohydrate scene. A detailed account of his life and scientific contributions is scheduled to appear in a later volume.

Washington, D.C. Derek Horton
January 2006

Christian Pedersen

Aleksander Zamojski

61. Volume 61

Editor's Preface.—The sixty preceding volumes in this Series have recorded significant advances in our knowledge of every aspect of the carbohydrate field. Some of the articles compile information of permanent reference value, while others have documented the interim state of knowledge in a developing area, drawing attention to gaps in our understanding, and providing pointers for future research. This current Volume 61 offers authoritative articles of both types, covering a wide variety of carbohydrate topics.

The field had its early beginnings in empirical studies on those macromolecular biomaterials we call carbohydrates, together with the study of simple sugars as organic compounds that present challenges of stereochemical complexity and multi-functionality. Over time the discipline has burgeoned, from its early bases in food and fiber technology and a specialized area of organic chemistry, onto center stage in its present wide role in chemistry, biochemistry, and glycobiology.

Mass spectrometry has had enormous impact on our ability to analyze carbohydrate structures. Its fundamentals for the analysis of sugars and their derivatives under electron impact were documented by Kochetkov and Chizhov in Volume 21 of this Series, and the advent of Fast Atom Bombardment significantly extended the technique to applications with molecules of considerably greater complexity, as recorded by Dell in Volume 45. Progress in instrumental capabilities in this rapidly evolving field has greatly augmented the reach of mass spectrometry in characterization of glycans and glycoconjugates, warranting an early revisit to the topic in the present article by Thomas-Oates and her team (York, UK). Interestingly, the traditional techniques for structural analysis of carbohydrates, namely permethylation, peracylation, and labeling of the reducing terminal, remain core methodologies within these mass-spectrometric applications.

Deoxy sugars are of wide occurrence as components of nucleic acids, natural glycosides, and antibiotics, and they were the subject of an earlier article by Hanessian in Volume 21. Here, de Lederkremer and Marino (Buenos Aires) provide a detailed update on the distribution of deoxy sugars in natural products, along with a survey of methods for their synthesis.

The most abundant simple carbohydrate, and indeed the most abundant pure organic chemical, is sucrose. Its central role as a nutritive sweetener is well documented, but its potential as a precursor for organic synthetic manipulation (sucrochemistry) has long been a story of largely unrealized potential. The article here by Queneau (Lyon), Jarosz (Warsaw), and their coauthors provides a comprehensive account of actual and potential uses of sucrose as a cheap precursor for a range of applications as food additives, pharmaceuticals, surfactants, and complexing agents. The authors assess fundamental questions of selective reactivity among the eight functional groups in the sucrose molecule, as well as practical aspects concerning processing conditions and reaction solvents.

The article by Islam and von Itzstein (Brisbane) focuses on the worldwide threat of a possible avian influenza pandemic. The key involvement in the infective process of sialidase, a surface glycoprotein of the influenza virus, provides a pointer to the design of effective drugs that can combat all pathogenic strains, rather than vaccines, which target a single strain. Inhibitors of the sialidase, based on the 1,2-unsaturated analogue (Neu5Ac2en) of N-acetylneuramininc acid have led via computational methodology to effective mimetics, most notably Relenza (zanamivir) and Tamiflu (oseltamivir), along with other inhibitors. The chapter details the very extensive synthetic and pharmacological background involved in reaching these optimized targets.

Nicotra's group in Milan contributed the article on chemoselective neoglycosylation, namely the chemical attachment of saccharide chains to various acceptors: proteins, peptides, lipids, and different types of support media, to encode specific carbohydrate determinants for molecular recognition. Synthetic realization of the specific N- or O-glycosyl links present in the glycoprotein structure is difficult, and other modes of ligation must be employed. The utility of such structures is manifestly evident in the clinical success of such conjugates as the *Haemophilus influenzae* type b (Hib) vaccine, the use of neoglycopeptides in cancer diagnosis, and the enhancement of protein-binding affinity by use of glycoarrays and dendrimers.

The development of carbohydrate science recorded in these volumes by noted experts during the past six decades has been closely paralleled by the content of conferences devoted to the field, where many of these specialists have signaled important progress. Most notably, the International Carbohydrate Symposia, beginning in the early 1960s and proceeding on a biennial basis, have reflected the remarkable growth and breadth of the field. The historical account of these symposia presented here by Angyal (Sydney), a frequent contributor to this Series, documents this evolution in detail. In addition to listing the principal themes and contributors, he offers valuable pointers for organizers of future symposia in the most effective use of these meetings for information exchange and the stimulation of new ideas.

The long and distinguished career of N. K. Kochetkov, a towering figure in Russian science, who made major contributions to the study of nucleotide sugars, the methodology of glycosidic coupling, and the mass spectrometry of carbohydrate derivatives, is recognized in the biographical memoir by Knirel and Kochetkova (Moscow).

Washington, D.C. Derek Horton
July 2007

Nikolai K. Kochetkov

62. Volume 62

Editor's Preface.—Sugars played a key role at the dawn of the nuclear magnetic resonance era. Foremost was Lemieux's demonstration that the magnitude of vicinal proton–proton spin couplings was related to the spatial orientation of the hydrogen atoms. Subsequently, Karplus presented an equation that quantifies the relationship between the NMR coupling constants of sugars and related molecules to the dihedral angle between the coupled nuclei. In this issue, Coxon (Bethesda, Maryland) provides a comprehensive survey of the many empirical variants of the Karplus-type equation that have been proposed over the years as NMR spectroscopy has evolved to become the indispensable key tool in carbohydrate characterization. The treatment covers spin couplings over three, four, and five bonds, and includes both protons and heteronuclei. The equations range from simple, two-parameter versions that depict only the torsional dependence of coupling constants, to complex 22-parameter forms that simulate the variation of the coupling with many different molecular properties. Earlier applications of NMR in the carbohydrate field were surveyed in this series by Hall in Volumes 19 and 29, by Coxon in Volume 27, and later by Csuk and Glänzer in Volume 46, and by Tvaroška and Taravel in Volume 51.

Three articles in this volume, written from different perspectives, focus on a central theme in carbohydrate science, namely the glycosidic linkage, its formation and cleavage. The presumed intermediates or transition states in these reactions have considerable oxacarbenium ion character. Using a computational approach, Whitfield (Ottawa) employs quantum mechanics to study the bonding characteristics of such ions, which are difficult to study experimentally.

Focusing on the 2,3,4,6-tetra-*O*-methyl-D-galactopyranosyl cation and its 2-*O*-acetyl analogue, his treatment examines their detailed conformations (half-chair and skew) and calculated energies to demonstrate the potential of *in silico* methodology in understanding the reactivity of glycopyranosyl oxacarbenium ions.

Although glycoscientists have learned over the years how to isolate certain classes of naturally occurring complex carbohydrates, the availability of pure natural compounds remains inadequate to address many ongoing challenges. Chemical synthesis remains the essential resource for accessing complex oligosaccharides and glycoconjugates, to provide significant quantities of the pure natural structures as well as unnatural mimetics that are often of interest.

In this volume, Smoot and Demchenko (St. Louis, Missouri) survey the drawbacks of traditional oligosaccharide syntheses that require extensive protecting-group manipulations between each glycosylation step, and discuss significant recent improvements that have emerged to circumvent the shortcomings of earlier approaches. Their article complements the classic chapter in Volume 50 by Schmidt and Kinzy on the trichloroacetimidate methodology, the articles by Garegg in Volumes 52 and 59 focusing on thioglycoside procedures, and solid-state synthetic techniques as presented in Volume 58 by Plante, Palmacci, and Seeberger.

In a third related article, Cai, Wu, and Crich (Detroit, Michigan) focus on the particular situation in homoglycan synthesis where the glycon has an axial hydroxyl group at the 2-position. For the total synthesis of α-mannans and α-rhamnans the overall strategy involves construction of the 1,2-*trans*-axial glycosidic bond, whereas the β-mannans and β-rhamnans require additional steps post-glycosylation to generate the 1,2-*cis*-equatorial glycosidic linkage.

The glycobiology of *Trypanosoma cruzi*, the causative agent of Chagas' disease, has contributed significantly to the identification of target enzymes responsible for the construction of unique cell-surface molecules. Agusti and de Lederkremer (Buenos Aires) here discuss the structure of glycoinositolphospholipids in *T. cruzi*, both free and as protein anchors, that display unusual structural motifs, including galactofuranose, which are absent in mammals. The chemical synthesis of oligosaccharides containing Gal*f* is presented, as well as the use of the organism's trans-sialidase (TcTS) for synthetic purposes.

Many of the articles in earlier volumes of this *Advances* series remain of permanent reference value, but library holdings of the full series have not kept pace with growth of the carbohydrate field and the opening of many new research centers. In an important development, the publishers have now made full-text electronic access available, through Science Direct, to all previously published articles in every volume. Volume 1, which appeared in 1945 with the title *Advances in Carbohydrate Chemistry*, under the editorship of Ward Pigman and Melville L. Wolfrom, featured as its opening article a chapter on the classic Fischer cyanohydrin reaction for ascent of the sugar chain, authored by the great carbohydrate pioneer Claude S. Hudson. The series title was changed with Volume 24 to *Advances in Carbohydrate Chemistry and Biochemistry* to reflect the broadening impact of carbohydrates in the biological field.

Grateful thanks are expressed to three senior statesmen of the carbohydrate community, Laurens Anderson, Hans Baer, and John Brimacombe, who as members of the Board of Advisors have helped over the years to guide these *Advances* with their own contributions, their valuable insight, and their mentoring service to others providing significant articles to the series.

With this issue, Robin Ferrier (New Zealand) and Mario Monteiro (Guelph, Ontario), the first and last doctoral students of the late Gerald Aspinall, pay tribute to the enormous contributions made by their mentor to the carbohydrate field, especially in the area of polysaccharide structural methodology. Roberto Rizzo (Trieste) provides an account of the life and work of Vittorio Crescenzi, and his notable investigations on polysaccharide technology and the physicochemical behavior of biopolymers in solution.

A sad note is of the recent deaths of several great carbohydrate pioneers of the twentieth century, including Roger Jeanloz, Bengt Lindberg, and Per Garegg, whose work will be commemorated in a future volume.

Washington, D.C. Derek Horton
November, 2008

Gerald O. Aspinall

Vittorio Crescenzi

63. Volume 63

Editor's Preface.—Synthesis has been a sustained area of major interest in carbohydrate science, and the preceding Volume 62 of this series featured three chapters focusing in detail on the construction of glycosidic linkages. Many established synthetic methods permit the elaboration of complex target molecules, but frequently involve the use of tedious protection–deprotection sequences and expensive or hazardous reagents. This current Volume 63 of *Advances* presents two articles that offer promise of useful methodologies for simplified procedures amenable to low-cost large-scale applications, using mild conditions and environmentally friendly materials.

Rauter and her coauthors Xavier, Lucas, and Santos (Lisbon) present here a detailed overview of the potential for heterogeneous catalysts in useful synthetic transformations of carbohydrates. Such silicon-based catalysts as zeolites are easy to handle and recover, are nontoxic, and can offer interesting possibilities for exercising stereo- and regio-control in many established carbohydrate transformations.

In the midst of wide-ranging research on the role of complex oligosaccharides in biological recognition processes, and the attendant focus on synthesis of such molecules, the important role of oligosaccharides in large-scale commercial applications is often overlooked. The contribution by Seibel and Buchholz (Wurzburg and Braunschweig) in this volume addresses in detail those tools of particular value for preparation of oligosaccharides that serve needs in the food, pharmaceutical, and cosmetic industries. Major emphasis is devoted to the use of readily available enzymes (glycosidases, glycosynthases, glucansucrases, fructansucrases) and abundant carbohydrate substrates, especially sucrose, and the applications of enzyme and substrate engineering. Particular attention is given to those large-scale applications of oligosaccharides that serve as sweeteners, as well as promising new medical applications.

Two complementary treatments deal with different aspects of the intense current interest in the biological recognition phenomena between carbohydrates and proteins. Dam and Brewer (New York) examine in detail the energetics and mechanisms of binding between lectins (carbohydrate-binding proteins) and the multivalent glycoprotein receptors on the surface of normal and transformed cells, as well as other types of carbohydrate receptors, including linear glycoproteins (mucins). The authors postulate a common "bind-and-jump" mechanism that involves enhanced entropic effects which facilitate binding and subsequent complex formation.

In an extensive and comprehensive survey, Chabre and Roy (Montreal) revisit the subject of neoglycoconjugates introduced three decades ago by Stowell and Lee in

Volume 37 of this Series. It was then a nascent topic, and the Montreal authors now bring together in a single large article the vast new literature base that has subsequently evolved in the field of the "glycoside cluster effect." The years have witnessed much creativity in the design and strategies of synthesis that have afforded a wide array of novel carbohydrate structures, and reflect the ongoing dynamic activity in this rapidly evolving area, even since the recent article by the Nicotra group in Volume 61 of this series. Elaborate nanostructures, now termed glycoclusters and glycodendrimers, feature arrays of carbohydrate epitopes joined via ligands onto a variety of scaffolds, including calixarenes, porphyrins, and such carbon nanostructures as fullerenes and nanotubes. Their synthesis and characterization are addressed in detail along with evaluation via such techniques as microarrays and other modern analytical techniques, for their potential in application to biological systems.

The contributions of two of the world's leading carbohydrate innovators are recognized in this issue. The work of Roger Jeanloz evolved from a classical background in synthesis and structure elucidation to wide applications in the biological area which have led him to be considered as the father of the subject now known as glycobiology. He contributed extensively to these *Advances,* with articles in Volumes 6, 11, 13, and 43 of the series, and his life and scientific work is the subject of the obituary article by Sharon (Israel) and Hughes (London).

The article by Kenne, Larm, and Alf Lindberg (Stockholm) surveys the career of Bengt Lindberg, and especially notes Lindberg's development of the methodology for microscale analysis of carbohydrate structures that has permitted determination of the sequence structure of minute samples of oligo- and poly-saccharides from biological sources and has enabled the explosive growth of modern research in glycobiology. Lindberg was also a notable contributor to this series, with articles in Volumes 15, 31, 33, and 48 that document in Lindberg's classic terse style the evolution of structural methodology from early beginnings to sophisticated applications, in particular for bacterial polysaccharides.

The death on October 8, 2009 of Antonio Gómez Sánchez is noted with regret. He was one of the successors of the Seville carbohydrate school built up by Francisco García González, with whom he coauthored in Volume 20 a chapter on the reaction of amino sugars with 1,3-dicarbonyl compounds, a subject that was a major theme of Gómez Sánchez's research.

Sincere thanks are expressed to Professors Stephen Angyal and J. Grant Buchanan for their many years of advice and support as members of the Board of Advisors. Welcomed as a new member of the Board is Professor Todd Lowary.

Washington, D.C.
November, 2009

Derek Horton

Roger W. Jeanloz

Bengt Lindberg

64. Volume 64

Editor's Preface.—The two most abundant organic substances on Earth, cellulose and starch, have been known since antiquity and have been put to many uses, although their chemical identity as polymers of D-glucose were not recognized until the early part of the 20th century. Remarkably, while our detailed knowledge of the structural and functional complexity of a great many carbohydrates in biological systems has burgeoned in recent years, many aspects of these two biopolymers have remained poorly understood. This current Volume 64 of *Advances* features two complementary chapters: Pérez and Samain (Grenoble) focus on the structure, morphology, and solubilization of cellulose, while Mischnick (Braunschweig) and Momcilovic (Stockholm) examine the chemical modification of cellulose and starch with the powerful newer analytical tools now available.

Fiber X-ray crystallography is the key tool used by the Grenoble group to focus on such organizational levels of cellulose structure as chain conformation, chain polarity, chain association, crystal polarity, and microfibril structure. Sundararajan and Marchessault, in Volume 36 of *Advances,* reviewed knowledge earlier gleaned by this technique, but long-standing questions have remained concerning the packing of chains in native cellulose in relation to cellulose modified by chemical treatment. Pérez and Samain now offer resolution of many of these questions and also provide a glimpse of promising new applications for cellulose based on the design of materials having unique physicochemical properties.

The chemical derivatization of cellulose and starch at the three available hydroxyl groups is a reaction of inherent high complexity on account of differences in the reactivity of the individual groups, questions of accessibility in heterogeneous systems, and their behavior of the groups under kinetic or thermodynamic conditions. Mischnick and Momcilovic take advantage of advanced spectroscopic and chromatographic methods to elevate the classic Spurlin model to a higher level of predictability for understanding the structure–properties relationships in chemically modified cellulose and starch. Such interpretations point the way for improving reproducibility and the rational design of products having desired properties. Much of the older work found in the patent literature on starch chemistry, as detailed in by Tomasík and coworkers in Volume 59 of this series, employed a largely empirical approach.

A key driving force behind much current research on carbohydrates has been the potential for applications in medicine and biology, as exemplified in the extensive chapter on neoglycoconjugates by Chabre and Roy in the previous volume (63) of this series. In the current volume, Marradi, Martín-Lomas, and Penadés (San Sebastián), working at the forefront of carbohydrate nanotechnology, detail the preparation and properties of nanoparticles functionalized with carbohydrates, with particular focus on gold glyconanoparticles wherein neoglycoconjugates bearing a thiol functional group are clustered around a gold nanoparticle to present a glycocalix surface. These carbohydrate-based nanoparticles have been used as tools in interaction studies with proteins in biological systems. The gold-based approach is further extended to the use of a variety of other materials, enabling access to multifunctional glyconanoparticles having a range of optical, electronic, mechanical, and magnetic properties.

The interaction of simple sugars with amines to initiate the series of transformations termed the Maillard reaction is well known in food chemistry, with its initial reaction to form a glycosylamine being succeeded by the Amadori "rearrangement" to afford a 1-amino-1-deoxyketose derivative. The Amadori rearrangement reaction between glucose and such biologically significant amines as proteins leads to an N-substituted 1-amino-1-deoxy-D-fructose (fructosamine) structure by the process termed "glycation." Since the 1980s, fructosamine has witnessed a boom in biomedical research, mainly due to its relevance to pathologies in diabetes, Alzheimer's disease, and related aging processes. In an extensive survey, Mossine and Mawhinney (Columbia, MO) detail important new knowledge on, and applications of, fructosamine-related molecules in basic chemistry, and in the food and health sciences.

Again reflecting medical and glycobiological aspects, research development on vertebrate sialidase biology during the past decade is addressed here by Monti (Brescia) along with coauthors Bonten (Memphis, TN), d'Azzo (Memphis, TN), Bresciani (Brescia), Venerando (Milan), Borsani (Brescia), Schauer (Kiel), and Tettamanti (Milan). Since Schauer's original chapter on sialic acids in Volume 54 of *Advances*, there have been many developments in our understanding of the important roles played by sialic acid-containing compounds in numerous physiological processes, including cell proliferation, apoptosis and differentiation, control of cell adhesion, immune surveillance, and clearance of plasma proteins. Within this context, sialidases, the glycohydrolases that remove the terminal sialic acid at the nonreducing end of various glycoconjugates, perform an equally pivotal function. The authors detail the subcellular distribution of these enzymes in mammalian tissues and their particular function, and also highlight the trans-sialidases, enzymes abundant in trypanosomes and expressing pathogenicity in humans.

Finally, this volume pays tribute to the life and work of two recently deceased carbohydrate scientists of note, Roy L. Whistler at the age of almost 98 and, younger by a whole generation, Per J. Garegg at the age of 75.

In a uniquely personal account, Whistler's student, James BeMiller, traces the threads of his mentor's long career in industry, government, and academia. Whistler's complex and ambitious character cast a powerful influence over the many who worked with him on the structure and applications of polysaccharides, where he was able to bridge effectively the academia–industry divide in the practical exploitation of numerous polysaccharides. Over the course of many decades, he was the key mover in the establishment and development of the international carbohydrate meetings, as detailed by Angyal in Volume 61 of *Advances*. He was also the author and editor of several books on polysaccharides and their industrial applications, as well as being a contributor to *Advances* and a member of its Board of Advisors.

In contrast, Per Garegg left his mark as a superbly talented synthetic carbohydrate chemist leading an enthusiastic and devoted group of coworkers in the construction of remarkably complex oligosaccharide sequences of relevance in medicine and biology. Working in Stockholm with his mentor Bengt Lindberg and later as Lindberg's successor, his synthetic targets focused on the carbohydrate sequences of bacterial polysaccharides whose structures had been elucidated by the Lindberg group. The tribute here by Oscarson (Dublin) and Larm (Stockholm), together with input from other friends and colleagues, paints a picture of a genial man much admired by his student coworkers, an important contributor to the *Advances*, and a familiar figure at the major carbohydrate meetings.

Washington, D.C.
July, 2010

Derek Horton

Roy Lester Whistler

Per Johan Garegg

65. Volume 65

Editor's Preface.—This volume pays tribute to the life and work of two notable figures in the carbohydrate scene, each of whom made major contributions in quite different areas. Jean Montreuil, who lived for almost ninety years, was a world pioneer in the field now known as glycobiology, and he was recognized as one of the great leaders in French biochemistry. He was a colorful and influential personality who held a lifetime attachment to the north of France and his native city of Lille, where he created the Lille Glycobiology School. The lively account presented here by Michalski, one of his associates in Lille, illustrates Montreuil's far-reaching scientific contributions, and his work in organizing many national and international conferences on glycoconjugates.

The career of Laurance David (Laurie) Hall took a quite different course; he was a mercurial figure who crisscrossed the Atlantic and played a major role in the development of NMR spectroscopy of sugars following on from the seminal work of Lemieux, and he later became a pioneer in what became known as magnetic resonance imaging (MRI). Setting out originally in synthetic carbohydrate chemistry in Hough's laboratory in Bristol, UK, he soon turned to the new field of NMR and made astonishingly prolific contributions in a career that developed in Vancouver at the University of British Columbia. Bruce Coxon (Bethesda, Maryland), his close friend from their days together in Bristol, and himself an NMR spectroscopist, traces in thorough technical detail the innovative work of Laurie Hall on the NMR of sugars bearing a wide range of different magnetic nuclei, and on to the last phase of Hall's work conducted back in the UK at Cambridge, where again he pioneered the use of magnetic resonance, especially MRI, in a wide range of physiological and technological applications. Indeed, had Laurie Hall not died at a relatively young age, the article here recorded by Coxon might have been a chapter written by Hall himself, as a sequel to his two earlier noteworthy contributions to *Advances*.

In the early days of this *Advances* series, authors were offered the opportunity to record experimental procedures as part of their contributions, but this has been rarely used in recent years. However, the practice is revived here in the article by El Nemr and El Ashry (Alexandria) on the synthesis of trehazolin, a natural nitrogen-containing pseudodisaccharide that is an inhibitor of the enzyme trehalase, together with various analogues. These synthetic targets can be approached in a wide variety of ways, and the authors present for each synthetic sequence sufficient detail to permit an experienced synthetic chemist to repeat the procedure.

The red algae are lower plants, mostly from marine sources, that contain a wide diversity of sulfated polysaccharides in their cell walls and intercellular matrices, and many of these have enjoyed long practical use as a consequence of their ability to form strong gels in aqueous solution. Such examples as agar and carrageenan are well known; their structures were surveyed by Mori in Volume 8 of this series and their physicochemical properties were the subject of a landmark article by Rees in Volume 24, but only with the advent of sophisticated new structural techniques has the vast range of different structures in these algae been recognized. Usov (Moscow) has been a leading contributor to the literature in this field, and his article in this volume provides a comprehensive overview of the known chemical structures, along with interesting new perspectives on the classification of the *Rhodophyta* according to the data of molecular genetics.

The terms glycomics and glycobiology are now firmly established as major themes of carbohydrate science, notably in relation to cell-surface glycoproteins and glycosphingolipids and their relation to human disease. The detailed analysis of glycans in these components offers high potential for medical/pharmaceutical applications. However, as Nishimura (Sapporo, Japan) points out in this volume, there are major problems involved in separating and analyzing the complex mixtures obtained from whole-serum glycoproteins, and this limits the current therapeutic potential. The author details the rapid evolution of automated methodology to overcome these difficulties, and offers promise for major advances in glycan profiling by employing the "glycoblotting" technique in conjunction with mass-spectrometric analyses.

John M. Webber, who died on January 4, 2011, was the author of an article on chitin in Volume 15 and higher-carbon sugars in Volume 17 of this series, and was a member of the editorial team that founded the journal *Carbohydrate Research*.

Nathan Sharon, a leading carbohydrate biochemist and long-time member of the Advisory Board of this series, died June 17, 2011. A detailed account of his life and work will appear in a future volume of *Advances*.

Washington, D.C. Derek Horton
May, 2011

Jean Montreuil

Laurance David Hall

66. Volume 66

Editor's Preface.—In this 66th volume of Advances, an extended tribute to the life and work of Anthony C. Richardson (Dick) is provided by Hale (Belfast). In the present day, when the successful young academic usually expects to abandon the laboratory bench and progress to become the leader of a large research group, Richardson was an anomaly in remaining an outstanding synthetic chemist who spent his entire career doing the laboratory work he loved. He had a brilliant and incisive mind coupled with a warm and nurturing personality, but he never sought the limelight or public recognition beyond the satisfaction from working with a small group and achieving a remarkable range of accomplishments in carbohydrate synthesis. These justly merit the extended documentation presented here in Hale's account. From elegantly conceived syntheses of (−)-swainsonine to Richardson's rules for predicting the outcome of sulfonate displacement by nucleophiles, to the noncaloric sucrose-derived sweetener Splenda® (trichloro-*galacto*-sucrose) with the Hough laboratory, and the provision of many valuable synthetic intermediates, he has enriched the carbohydrate field with a very substantial legacy.

The coordination behavior of sugars and their derivatives with inorganic cations has been largely "under the radar" of mainstream carbohydrate science in recent years, given the strong focus of many of today's researchers on glycobiology targets. However, the complexation of carbohydrate derivatives with the element chromium, in particular, has important implications in both human and animal health, and in problems of environmental damage from industrial pollutants. The toxicity and carcinogenicity of chromium is well recognized, and the use of microorganisms or plants for bioremediation of contaminated soils requires careful evaluation. The unpaired d-subshell electrons in the multiple valence states exhibited by chromium lend themselves ideally to studies of the complexes by electron paramagnetic resonance. This chapter by Sala and colleagues (Rosario, Argentina) details current knowledge gleaned from use of traditional continuous-wave EPR spectrometers and addresses the potential of newer pulsed and high-field instruments for significant advancement of our understanding.

While the pyranose and furanose ring forms of the sugars dominate the carbohydrate literature, the uncommon septanose ring forms have long intrigued sugar chemists, with particularly notable contributions by Stevens in Australia. The chapter in this volume from Saha and Peczuh (Storrs, Connecticut) provides a comprehensive overview of the subject from both the synthetic and structural viewpoints, and presents a detailed analysis of the conformational behavior of these ring forms. It may be noted that the structures of the seven-membered rings and their acyclic precursors are most conveniently depicted, respectively, by Mills-type formulas and the supposedly old-fashioned Fischer-type formulas, rather than the Haworth conformational formulas commonly favored for five- and six-membered rings. This chapter offers intriguing prospects for involvement of these ring forms in biological applications, especially with regard to antisense oligonucleotides.

A most important variant of the monosaccharide structures is the class of sugar derivatives wherein nitrogen replaces the ring-oxygen atom, namely, the imino sugars. In the comprehensive overview presented here by Stütz and Wrodnigg (Graz, Austria), they integrate the wide range of imino sugar analogues now known to occur in Nature with their remarkable functions as potent inhibitors of the glycosidase enzymes. Complementary work on numerous synthetic analogues has added important new understanding of the mode of action of glycosidases in general. The authors include a detailed structural tabulation of all such inhibitors currently known, along with the Protein Data Bank links to the enzymes that they inhibit, and offer exciting prospects for the therapeutic potential of these inhibitors in modulating essential metabolic processes.

The deaths are noted with regret of two leading carbohydrate biochemists, Nathan Sharon (June 17, 2011) and Saul Roseman (July 2, 2011). Dr. Sharon served with distinction for many years as a member of the Board of Advisors of this *Advances* series, and his advice and input will be sadly missed. His work will be recognized in an upcoming volume of the series.

With this volume Professor Arnold E. Stütz is welcomed as a new member of the Board of Advisors.

Washington, D.C. Derek Horton
October 2011

Anthony (Dick) C. Richardson

67. Volume 67

Editor's Preface.—Crystallographic structure-determination of sugars was pioneered by George Jeffrey in his classic 1939 study of the monosaccharide glucosamine, which showed definitively the $^4C_1$ conformation of the pyranose ring. For polysaccharides, and in particular, the most abundant and widely used carbohydrate, namely, cellulose, a key advance in our knowledge was the 1937 crystallographic study by Meyer and Misch. Their work, building from 1926 proposals by Sponsler and Dore, demonstrated a plausible unit cell for this biopolymer and showed that cellulose is a (1→4)-linked polymer of glucopyranose residues. However, many detailed aspects of the structure of cellulose remain enigmatic even to this day.

The present article by French (New Orleans) addresses questions of the fine structure of cellulose by drawing on a combination of crystallographic methodology and computational chemistry, utilizing, in particular, the three-dimensional structure of various cellobiose derivatives. French's contribution complements the recent articles in Volume 64 by Pérez and Samain focusing on structure and engineering of cellulose and that of Mischnik and Momcilovic on chemical derivatization of cellulose, as well as earlier work on cellulose detailed by Marchessault and Sundararajan in Volume 36, and by Jones in Volume 19.

The glycosaminoglycan heparin is a very complex carbohydrate biopolymer that has long been pharmacologically important as an anticoagulant. Its isolation from tissue sources and its properties were first surveyed in this series by Foster and Huggard in Volume 10. Our understanding of its detailed structure and manifold biological functions has presented many challenges, as documented by Casu in Volume 43 and by Casu and Lindahl in Volume 57, with particular emphasis on the pentasaccharide subunit that binds to antithrombin III in a key step of the coagulation process. Major advances in synthetic strategy in recent years have enabled the chemical synthesis of a pentasaccharide drug, fondaparinux (Arixstra®), that is clinically effective for the treatment of deep-vein thrombosis. Dulaney and Huang (East Lansing, Michigan) present in this volume a detailed comparative overview of strategies, reported during the most recent decade, for the chemical synthesis of related pentasaccharide subunits of heparin and heparan sulfate, including both solution-phase and solid-support approaches, the active–latent strategy, and chemoenzymatic methodology.

Another area where synthetic carbohydrate chemists have displayed spectacular virtuosity is in the construction of the glycosylphosphatidylinositol (GPI) glycolipids that anchor cell-surface proteins and glycoproteins to the plasma membrane of eukaryotic cells. Earlier work on the GPI anchor of the protozoan parasite *Trypanosoma cruzi* was described by Lederkremer and Agusti in Volume 62. In this volume, Swarts and Guo (Detroit, Michigan) survey the intensive current efforts of research groups, including their own and by such noted synthetic investigators as Ogawa, Schmidt, Fraser-Reid, Ley, Seeberger, Vishwakawa, and Nikolaev, to synthesize a range of intact GPI anchors. These anchors play important roles in many biological and pathological events, including cell recognition and adhesion, signal transduction, host defense, and acting as receptors for viruses and toxins. Chemical synthesis of structurally defined GPI anchors and related analogues is a key step toward understanding the properties and functions of these molecules in biological systems and exploring their potential therapeutic applications.

The final article, by Serpersu and Norris (Knoxville, Tennessee), surveys dynamic and thermodynamic aspects of the interactions between various aminoglycoside-modifying enzymes (AGMEs) and their antibiotic ligands. It complements the fundamental articles in Volume 30 by the Umezawa brothers on the structure of the aminoglycoside antibiotics and the mechanisms whereby mutant organisms resist such antibiotics, along with the studies reported in Volume 60 by Willis and Arya on aminoglycoside–nucleic acid recognition. The Tennessee authors stress the key role of two important aminoglycoside-modifying enzymes in their complexation with the aminoglycosides, as studied by the thermodynamics of ligand–protein interactions and the dynamic properties of the protein.

A tribute to the late Nathan Sharon is presented by three of his students and colleagues, David Mirelman, Edward A. Bayer, and Yair Reisner (Rehovot, Israel). Sharon was a major pioneer in carbohydrate biochemistry, and his outstanding contributions to our knowledge of lectins are particularly noteworthy. He was a long-time member of the Board of Advisors of this series.

Stephen J. Angyal, for many years a member of the Board of Advisors and a noted expert on carbohydrate stereochemistry, died on May 14, 2012, and Hassan S. El Khadem, a frequent contributor to this series on nitrogen heterocycles derived from sugars, passed away on May 20, 2012.

Washington, D.C. Derek Horton
April 2012

Nathan Sharon

68. Volume 68

Editor's Preface.—This volume of *Advances* constitutes a departure from the customary format in that it features one article on a rapidly evolving new field of research in its early stage of development, together with another very large article that presents a broad overview of a mature area of carbohydrate science.

The contribution from Nakagawa and Ito (Saitama, Japan) focuses on small molecules that mimic the carbohydrate-binding properties of lectins, a class of proteins commonly derived from plant sources and which are used extensively as research tools in glycobiology. Lectins offer interesting potential in drug applications but are expensive to manufacture, have low chemical stability, and may be immunogenic.

Consequently, there is high current interest in molecules that behave as lectin mimics, emulating the strong carbohydrate-binding properties of natural lectins. Some of these occur naturally (pradimicins and benanomicins) and others are readily accessible by synthesis.

The synthetic examples fall into two broad categories, the boronic acid-dependent and boronic acid-independent lectin mimics, and they lend themselves readily to chemical modification for "tuning" the architecture of the carbohydrate-binding cavity and optimizing the strength of binding.

Although the carbohydrate-binding ability of most of the lectin mimics studied thus far does not reach the level of natural lectins, their therapeutic potential as antiviral and antimicrobial agents clearly warrants active pursuit toward goals in medicine. In the broader context, these studies will further our understanding of the molecular basis of carbohydrate recognition.

The most abundant carbohydrates in the biosphere are cellulose, chitin, and starch. A recorded procedure for the isolation of starch from cereal grains, by the forerunner of today's wet-milling process, dates back two millennia to the writings of Pliny The Elder. The manifold applications to which starch has been put during those 2000 years, by conversion processes involving enzymatic, chemical, and physical transformations, long predate most modern scientific documentation.

In previous volumes of *Advances,* Tomasík (Cracow) has contributed comprehensive surveys of the conversions of starch by chemical and physical methods. The huge volume of work on the enzymatic transformations of starch published during the past 200 years, much of it the subject of claims in the patent literature, greatly exceeds the usual scope of an article in *Advances,* but there has been no obvious way to subdivide this body of work into a series of shorter articles. The large report contributed here by

Tomasík, in conjunction with this writer (Horton), surveys methods for the enzymatic conversion of starch by hydrolases and other enzymes, together with the role of microorganisms producing such enzymes and applications of these enzymes in the food, pharmaceutical, pulp, textile, and other branches of industry.

An effort has been made to coordinate the descriptions of the enzymes and their action patterns within the framework of the modern EC classification system. However, much early work, as well as some contemporary technology, has involved the use of crude enzyme isolates that are mixtures of individual enzymes, along with starches from different plant sources that exhibit distinctive differences in their behavior toward enzymes.

The two largest classes of starch-degrading enzymes are named uniformly in this article as alpha amylases and beta amylases. Although the terms α-amylase and β-amylase are very frequently encountered in the literature, the Greek designators are properly reserved to designate stereochemical configuration at the anomeric center in a formal sugar name. Applying the Greek designators to the enzymes has led some instructors and students to the erroneous notion that α-amylases cleave alpha linkages and β-amylases cleave beta linkages.

The products of enzyme-catalyzed starch degradation, mixtures of variously depolymerized glucosaccharides, are frequently described by the ill-defined term "dextrins," and this term may have a multiplicity of meanings in different situations. Likewise, many different names have been used for various enzymes and their plant, fungal, and microbial sources. As part of a comprehensive treatment, the article incorporates much information on starch enzymology recorded in the patent literature. Such sources have clearly not been subject to the peer-review process of the major journals, and they must be judged in this light; inconsistencies and contradictions remain evident.

Future issues of *Advances* are expected to revert to the regular format of five or six individual articles on a variety of carbohydrate topics.

The death is noted with regret of Malcolm B. Perry, on June 25, 2012. Dr. Perry worked for many years in Ottawa at the National Research Council of Canada, where he made extensive contributions in the carbohydrate field, especially on the structures of lipopolysaccharide O-antigens of gram-negative bacteria.

Washington, D.C.   Derek Horton
October, 2012

69. Volume 69

Editor's Preface.—This 69th volume of *Advances* features six contributions that cover a wide diversity of different aspects of new developments in the carbohydrate field, focusing variously on synthetic methodology, structural and functional aspects of carbohydrates in the bacterial cell-envelope, and a glycobiology theme concerned with aberrant glycosylation on the surface of tumor cells and approaches to improved cancer therapies.

The Boston (Massachusetts)-based group led by O'Doherty and his coworkers Aljahdali, Shi, and Zhong provide a tour-de-force of organic synthetic virtuosity in the *de novo* synthesis of a wide range of monosaccharides and oligosaccharides, making use of newly introduced asymmetric catalysts to accomplish high enantiomeric purity in the products.

Early work by Lespieau in Volume 2 of this series, and notably by Zamojski in Volume 34 provided pointers for the synthesis of alditols and aldoses "from scratch" employing simple organic precursors, but the methodology led mostly to racemic products. Important later work detailed here by the Boston group has focused on access to enantiopure products, as exemplified by the asymmetric epoxidation strategy of Sharpless, and complementary work by Danishefsky, MacMillan, Dondoni, Seeberger, and others.

The O'Doherty group has synthesized an impressive series of targets by employing an extensive range of "name" reactions, specialized reagents, "fine-tuned" versions of traditional oxidative and reductive procedures, and in particular the rearrangement reaction of enones introduced by Achmatowicz, utilizing asymmetric catalysts, along with numerous applications of palladium chemistry. These targets include the *C*-glycosyl framework of the papulacandins, the nojirimycin-type imino sugars, homoadenosine analogues, the pheromone daumone, the indolizidine alkaloids, analogues of trehalose and cyclitols, cardiac glycosides, the anthrax tetrasaccharide, and other carbohydrate structures of current biological interest.

The Vienna-based authors Messner, Schäffer, and Kosma detail the glycoconjugates of the cell envelope of bacterial and archaeal organisms, where glycosylation plays a critical role in preserving the integrity of the prokaryotic cell-wall and serves as a flexible adaption mechanism to evade environmental and host-induced pressure. Glycosylation is not limited to surface-layer proteins in the cell, but is shown to occur also in pili, flagella, and in the form of secondary cell-wall polysaccharides. In contrast to the glycan components in glycoproteins of eukaryotic organisms, these prokaryotic-associated glycans display a wide range of structural variations. With their incorporation of numerous unusual monosaccharide components, these glycans manifest a diversity in their constituent monosaccharides comparable to the complex range of sugar components found in the lipopolysaccharides and capsular polysaccharides of bacteria.

The discussion of these conjugates covers a range of aspects, from analysis and structure elucidation to function, biosynthesis, and genetic basis, and to biomedical and biotechnological applications. The S-layer glycoproteins form ordered structures through crystallization of the protein component, and a variety of techniques have helped to develop models for the structure of the cell envelope. Typically the glycan component consists of repeated complex linear or branched oligo- or polysaccharides, either O- or N-glycosylically linked, and the highly diversified structures have been elucidated by chemical and spectroscopic procedures. Their biosynthesis via the nucleotide-sugar pathway, and elucidation of their genetic basis, point the way to important applications in nanotechnology and biomedicine.

The article by Hevey and Ling (Calgary, Alberta) is focused on inhibitors designed to attenuate the metastasis of tumors through interruption of interaction between cell-surface lectins (carbohydrate-binding proteins) and those carbohydrate ligands (notably TF, sialyl Tn, and sialyl Le^x) that are typically overexpressed on tumor cells. These lectin–ligand interactions are correlated with such metastatic processes as cell adhesion, generation of new blood vessels (neoangiogenesis), and immune-cell evasion. The suppression of such interactions is a desired goal in seeking improved cancer therapies.

The authors discuss the structural features of the carbohydrate ligands in the tumor cell and the relation between abnormal glycosylation and tumor metastasis, and the role of lectins in tumor development. Focusing on the particular lectins involved, namely the galectins, selectins, and the siglecs (sialic acid Ig-like lectins), their structures and binding-sites are developed in detail.

The key role of the carbohydrate–lectin interaction suggests a variety of approaches toward novel cancer therapies. They include the development of inhibitors of the carbohydrate–ligand interaction, the inhibition of glycosyltransferases and glycosidases involved in the aberrant glycosylation of tumor cells, and the targeting of the immune system against cancer cells. Although there have been promising results from clinical trials, especially from dual-mode therapies, as compared to traditional cancer chemotherapy agents, there remain significant challenges to be addressed, with a need for carbohydrate inhibitors that are more resistant to hydrolytic or metabolic breakdown. Promising results have been observed by the use of small-molecule inhibitors based on C-glycosyl structures, which are resistant to acid hydrolysis and glycosylase enzymes, and polyvalent ligands that have shown enhanced binding down to the nanomolar level.

The contributions to carbohydrate science of three towering figures in the field are recognized in this issue with obituary articles that detail their work in sufficient depth to provide a valuable source for citation. Stephen J. Angyal, who died at the age of 97, was a Hungarian-born Australian who literally "wrote the book" on the chemistry of the cyclitols, developed a quantitative understanding of the conformational behavior of pyranoses in solution, and provided a wealth of practical synthetic procedures based on use of cationic complexes to modify the outcome of traditional reactions. His legacy is recorded here by his colleague John D. Stevens.

The scientific career of J. Grant Buchanan is presented by his colleague Richard Wightman. Grant's early studies in Alexander Todd's group in Cambridge provided a basis for his long-term interest in sugar phosphates and nucleotide chemistry, which developed further in his postdoctoral work with Calvin in Berkeley and later back in Britain at posts in London and in Newcastle. In conjunction with Baddiley, he studied the pneumococcal antigens and the ribitol teichoic acids. He also made major contributions to our understanding of epoxide migration and cleavage. In a subsequent move to his native Scotland, to Heriot–Watt University in Edinburgh, he extended his research to the chemistry and biochemistry of C-nucleosides, and the use of glycosylalkynes in synthesis. The list of his publications details his wide range of research accomplishments.

Saul Roseman was a pioneer American glycobiologist whose six decades of innovations in the field are documented by his student and coworker Subhash Basu. Very early in his career, Roseman established the correct structure of neuraminic acid (sialic acid), and subsequently he made extensive contributions to our knowledge of glycoproteins and glycosphingolipids, and the glycosyltransferases involved in their biosynthesis. Critical to this work was the availability of ^{14}C-labeled nucleotide sugars produced in collaboration with Khorana in Vancouver. Roseman's groundbreaking work on the glycosyltransferases set the stage for development of this theme in many centers of research in many centers around the world.

The deaths are noted of several past contributors to this series. William George Overend (died 18 September 2012, aged 90) of Birkbeck College, University of London, authored with Maurice Stacey an article on Deoxy Sugars in Volume 8, and with Finch wrote the obituary of Stacey in Volume 52. Robin Ferrier (after whom are named two "Ferrier Reactions"), of Victoria University, Wellington, New Zealand, contributed articles on Unsaturated Sugars in Volumes 24 and 58, and one on Boronates in Volume 35; he died July 11, 2013, aged 81. Serge David of the University of Paris authored with Estramareix an article in Volume 52 on Sugars and Nucleotides and the Biosynthesis of Thiamine. He died August 1, 2013, aged 92.

With this issue Jésus Jiménez-Barbero is welcomed as a member of the Board of Advisors.

Washington, D.C. Derek Horton
August, 2013

Stephen J. Angyal

PROFESSOR JOHN GRANT BUCHANAN

J. Grant Buchanan

Saul Roseman

70. Volume 70

Editor's Preface.—This 70th volume of *Advances* constitutes a small supplementary issue that pauses to review the evolution of the series since it began under the title *Advances in Carbohydrate Chemistry* toward the close of the Second World War, and looks toward its future as *Advances in Carbohydrate Chemistry and Biochemistry* in the ongoing documentation of a vastly expanded field of scientific disciplines where carbohydrates play a role.

The volume opens with an appreciation of the career of Struther Arnott and his contributions to our understanding of polysaccharide structure by use of X-ray diffraction studies of oriented fibers. The celebrated work of Crick and Watson, unraveling the structure and anticipating the all-encompassing biological role of DNA, was based on the X-ray diffraction studies by Maurice Wilkins and Rosalind Franklin at King's College, London. Arnott's later work in Wilkins' laboratory greatly extended our knowledge and understanding of the three-dimensional architecture of a wide range of polynucleotide chains by application of his Linked-Atom Least-Squares (LALS) methodology. Arnott's biographer, Rengaswami Chandrasekaran, details his subsequent researches at Purdue University in the USA that extend the compass of the technique to the structural characterization of a broad range of polysaccharides having relevance in both biological and technological areas. Equipped with great research talent, coupled with high administrative capacity, Struther Arnott fulfilled important leadership roles at Purdue and subsequently at St. Andrews University in his native Scotland. The Bibliography appended to the memoir provides a comprehensive source of reference to the many polysaccharide structures expertly documented by the Arnott research teams.

Struther Arnott

The *Advances* series set out in 1944 with the objective of presenting critical and integrating reports, understandable by the general reader as well as the specialist, on a wide range of topics having carbohydrates as a common theme. The report "Seven Decades of Advances" in this current volume assesses broadly the literature record on carbohydrates, as documented in over 350 articles published in this series during the seventy years of its existence, and its relation to research papers published in primary journals, as well as information in reference books, monographs, and text books that constitute the secondary literature.

At the time this series began, reports on original research were being published in a wide range of national journals and in many different languages. Some papers described sound, original, and novel work, but others may have had errors of fact, be of dubious originality, or had failed to give credit to relevant prior work. Although the more-prestigious national journals exercised a level of quality control, the peer-review system had not then developed to its current broad extent. In areas such as the carbohydrates, where a uniform system of nomenclature did not exist, there was a great deal of confusion in naming compounds and in their structural depiction.

It was the goal of a small group of leading carbohydrate chemists, notably Melville L. Wolfrom and Claude S. Hudson in the USA, and Sir Norman Haworth in Great Britain, and with the strong support of a far-sighted publisher, Kurt Jacoby (who had just founded Academic Press), to produce an annual book series featuring reports from invited experts on a wide distribution of topics related to carbohydrates. These areas would include simple sugars, glycosides, oligosaccharides, and polysaccharides, and deal with their structures, chemical and biochemical reactions, analytical chemistry, food and fiber technology, and other aspects.

Their endeavors set the stage from which the 70 volumes of *Advances* subsequently developed during a span of seven decades, averaging one volume per year. The authors contributing to the early volumes came mainly from North America and Great Britain, but the sources have expanded progressively to include articles from authors in countries throughout the world, from Europe, Australia, New Zealand, South America, Asia, and Africa. During this period, Academic Press took a prominent position in documenting other aspects of the carbohydrate field, with such series as Whistler's *Methods* and Aspinall's *Polysaccharides* volumes, and the Pigman books. When Academic Press became part of the Elsevier line, the breadth was further enhanced with the founding of the journal *Carbohydrate Research*.

The articles published in *Advances* have adhered to the original policies of its founders, and many of them remain definitive treatments of their subject, while others reflect advances in areas still in the course of rapid development. Collected here in chronological sequence are the published Prefaces to the individual volumes, and these describe the subject material in all of the individual articles, as well as providing contemporary commentary at the time the volumes were published.

Parallel with the evolution of the series there has been important progress in defining the language of scientific communication through agreements on standardized nomenclature; this has been regularly documented in the pages of the *Advances* volumes. It is a tribute to the cooperative efforts volunteered by a number of individuals that the standardization reflected in the volumes has served as a benchmark for widespread adoption of these nomenclature recommendations by authors throughout the world.

An important feature of *Advances* has been regular obituary reports that document the lives and motivations of major figures who have made significant contributions to the carbohydrate field. For the particular benefit of researchers who may not have had the opportunity to meet or hear these past leaders in person, this volume features a collection of portraits of all of those past leaders whose stories have been told in the various volumes.

The editorship of the series was, with one brief interruption, in the hands of its prime motivator, M. L. Wolfrom, until his death in 1969. Robert Stuart Tipson shared the editorship with Wolfrom, starting with Volume 8 in 1954. He worked jointly with the present editor until Volume 48 (1990), with the latter continuing thereafter.

The initiative of Academic Press to make back issues of *Advances* accessible electronically was significantly enhanced when the Elsevier organization established electronic access to a wide swath of the literature on carbohydrates through Science Direct, Scopus, and other sources. The massive chemical database of *Chemical Abstracts* provides records on individual carbohydrates and their derivatives, estimated to comprise at least 5% of the 73 million or more substances in the CAS Registry, and all are accessible electronically through SciFinder.

A bibliometric analysis of *Advances*, made in early 2013 by Professor Todd Lowary, revealed that 68 volumes of the series since the first issue had a total of 358 individual articles, carrying altogether 24,558 literature citations. These reports were cited 20,582 times, with an average of 70 citations per article and a 2012 impact factor of 7.133. The five most-cited reports were Bock and Pedersen's ^{13}C-NMR of monosaccharides article in Volume 41, with 1303 citations, the Schauer report on sialic acids in Volume 40 (858 citations), the Volume 41 ^{1}H-NMR glycoprotein article by Vliegenthart, Dorland, and van Halbeek (843 citations), the Schmidt–Kinzy trichloroacetimidate methodology in Volume 50, cited 694 times, and Legler's article in Volume 48 on the mechanism of glycoside hydrolases, which had 593 citations.

The content of this *Advances* series provides a large and unified source of reliable reference material contributed by noted experts in the field and is presented in clear, understandable English without unnecessary jargon or incomprehensible acronyms. The excellent support of the contributing authors and members of the Board of Advisors is expected to assure a fruitful ongoing continuation of the series in the service of the carbohydrate community.

Washington, D.C.
Columbus, Ohio, September 2013

Derek Horton

III. Concluding Remarks

The success and longevity of *Advances* owes much to the wisdom of its founders, most notably Melville L. Wolfrom, and the advice in particular of Claude S. Hudson and Sir Norman Haworth. They established, toward the close of the Second World War, a transatlantic scientific alliance to produce an annual book series of invited reports that would coordinate new work over a wide range of formal scientific and technical disciplines that share carbohydrates as a common theme. They were fortunate to have the support and foresight of Kurt Jacoby as he developed his fledgling scientific publishing house, Academic Press, and the expertise of members of an Executive Committee (later Board of Advisors) drawn from leaders of the time in academia and industry, especially Ward Pigman and Sidney Cantor.

The founders established policies for the series that required contributed articles to be critical and integrating and not mere literature reviews. Authors were expected to pay careful attention to historical accuracy and completeness, and write in clear English readily understandable by the average chemist or biochemist and not just the specialist. Such articles were designed to facilitate communication between scientists working in a variety of different disciplines that shared the mutual theme of carbohydrates, and efforts to establish a common language for naming carbohydrates was an important goal.

These basic policies have been sustained over the many years that *Advances* has been in existence, and have served as a force for stability and open communication in a field where local jargon, arbitrarily coined in specialized disciplines, has long impeded communication and understanding. At the time the series began, Professor Wolfrom himself had brought major initiative toward systematizing the naming of carbohydrates through the work of a committee of the American Chemical Society that he headed, but as he pointed out on several occasions, carbohydrate nomenclature remained a persistent problem. His leadership played an important role in securing a common understanding, first with British chemists and later with broader international agreement, but at the time of his death in 1969 he still considered that the systems under discussion at that stage remained a "work in progress."

Wolfrom would have been pleased with the subsequent broad international agreement reached in the 1996 IUPAC document *Nomenclature of Carbohydrates*[6] which provided a unifying basis for naming a wide range of simple sugars, their derivatives, oligosaccharides, polysaccharides, and various glycoconjugates. It provided directives for naming and depicting cyclic sugars and their conformations, stereochemical relationships, a three-letter code system for abbreviated depiction of saccharide sequences, and many other situations. Such arbitrary or nonstandard names as "acetobromoglucose," "diacetoneglucose " "pseudo sugar," or "azasugar" are now replaced by rational names. The consistent policy of *Advances* in adherence to these norms have stimulated wide acceptance, and have served to stabilize broad agreement and understanding between carbohydrate scientists.

Except for a few of the early volumes, Professor Wolfrom served as the Editor for *Advances* from its inception in 1944 until he died in 1969. Robert Stuart Tipson, a graduate from Haworth's Birmingham laboratory in England, joined him as coeditor starting with Volume 8 and continued until Volume 48, the year before his death in 1991. The present writer, Derek Horton, assisted Wolfrom with issues starting with Volume 14, becoming coeditor with Tipson starting in 1969 with Volume 24, and has subsequently continued as sole editor with Volume 25 and on, up to the current Volume 70.

It is hoped that this retrospect will serve as a coordinated source of information on all areas of carbohydrate science and the people who have played key roles in the development of the field.

For more than fifty years, many people in the worldwide carbohydrate community have enjoyed knowing my wife, Dr. June Horton, meeting as a friend at many national and international conferences, as a hostess at the Horton home, an appreciative guest on travels around the world, and as a friend and mentor to numerous graduate students and postdoctoral researchers in the Horton carbohydrate group. She lost her battle with cancer on November 10, 2013, as this issue was going to press.

References

1. E. F. Beschler, Walter J. Johnson and Kurt Jacoby: Academic Press, in R. Abel and G. Graham, (Eds.) *Immigrant Publishers: The Impact of Expatriate Publishers in Britain and America in the 20th Century,* Transaction Publishers, Rutgers—The State University, New Jersey, 2009, pp. 69–88.
2. D. Horton, Development of carbohydrate nomenclature, in K. L. Loening, (Ed.), *The Terminology of Biotechnology: A Multidisciplinary Problem,* Springer-Verlag, Berlin, 1990, pp. 41–49.
3. Rules of carbohydrate nomenclature, *Chem. Eng. News,* 26 (1948) 1623–1628.
4. Rules of carbohydrate nomenclature, *J. Chem. Soc.,* (1952) 5108; *Chem. Eng. News,* 31 (1953) 1776–1783.
5. Rules of carbohydrate nomenclature, *J. Org. Chem.,* 28 (1963) 281–291.
6. IUPAC Commission on the Nomenclature of Organic Chemistry (CNOC) and IUPAC–IUB Commission on Biochemical Nomenclature (CBN), Tentative rules for carbohydrate nomenclature, Part 1, 1969, *Biochem. J.,* 125 (1971) 673–695.
7. IUPAC–IUBMB Joint Commission on Biochemical Nomenclature, Nomenclature of Carbohydrates (Recommendations 1996), *Pure Appl. Chem.,* 68 (1996) 1919–2008; IUPAC–IUBMB Nomenclature of carbohydrates. *Adv. Carbohydr. Chem. Biochem.,* 52 (1997) 43–177.
8. Nomenclature of Carbohydrates, http:/www.chem.qmul.ac.uk/iupac/2carb/.

9. IUPAC –IUBMB Nomenclature of Glycolipids (Recommendations 1997), *Pure Appl. Chem.*, 69 (1997) 2475–2487; *Adv. Carbohydr. Chem. Biochem.*, 55 (1999) 311–326; http:www.chem.qmw.ac.uk/misc/glylp.html.
10. W. W. Pigman and R. M. Goepp, *Chemistry of the Carbohydrates*, Academic Press, New York, 1948, 748 pp.
11. W. W. Pigman, *The Carbohydrates: Chemistry, Biochemistry, Physiology*, Academic Press, New York, 1957, 902 pp.
12. W. Pigman and D. Horton, (Eds.) *The Carbohydrates, Chemistry and Biochemistry*, 2nd edn. Academic Press, New York, 1972, Vols. IA, IIA, IIB; 1980, Vol. IB.
13. R. L. Whistler and M. L. Wolfrom, (Eds.) *Methods in Carbohydrate Chemistry*, Volume 1, Academic Press, New York, 1962; Volume 2, 1963; Volume 3, 1963; Volume 4, 1965; R. L. Whistler and J. N. BeMiller (Eds.), Volume 5, 1965; Volume 6, 1972; Volume 7, 1976; Volume 8, 1980.
14. G. O. Aspinall, (Ed.), *The Polysaccharides*, Vol. 1, Academic Press, New York, 1982; Vol. 2. 1983; Vol. 3, 1985.
15. D. Horton, The development of carbohydrate chemistry and biology, in H. G. Garg, M. K. Cowman, and C. A. Hales, (Eds.), *Carbohydrate Chemistry, Biology and Medical Applications*, Elsevier, Amsterdam, 2008, pp. 1–28.
16. M. Pinto, (Ed.), *Comprehensive Natural Products Chemistry: Carbohydrates and Their Derivatives, Including Tannins, Cellulose, and Related Lignins*, Vol. 3, Pergamon Elsevier, Oxford, 1999.
17. S. Dumitriu, (Ed.), *Polysaccharides: Structural Diversity and Functional Versatility*, 2nd edn. Marcel Dekker, New York, 2005, 1176 pp.
18. B. O. Fraser-Reid, K. Tatsuta, J. Thiem, G. L. Côté, S. Flitsch, Y. Ito, H. Kondo, S. Nishimura, and B. Yu, (Eds.), *Glycoscience, Chemistry and Chemical Biology*, 2nd edn. Springer, Berlin, 2008, 2874 pp.
19. J. P. Kamerling, G. J. Boons, Y. C. Lee, A. Suzuki, N. Taniguchi, and A. G. J. Voragen, (Eds.), *Comprehensive Glycoscience*, Vols.1–4, Elsevier, Amsterdam, 2007.
20. P. M. Collins, (Ed.), *Dictionary of Carbohydrates*, 2nd edn. Chapman & Hall/CRC, Boca Raton, FL, 2006, 1282 pp.
21. R. Cammack, T. K. Attwood, P. N. Campbell, J. H. Parish, A. D. Smith, J. L. Stirling, and F. Vella, (Eds.), *Oxford Dictionary of Biochemistry and Molecular Biology*, 2nd edn. Oxford University Press, London, 2006, 720 pp.
22. R. D. Guthrie and J. Honeyman, *Introduction to Carbohydrate Chemistry*, 4th edn. Oxford University Press, London, 1974, 120 pp.
23. P. M. Collins and R. Ferrier, *Monosaccharides: Their Chemistry and Their Roles in Natural Products*, Wiley, Chichester, UK, 1995, 574 pp.
24. R. V. Stick and S. J. Williams, *Carbohydrates, The Essential Molecules of Life*, 2nd edn. Elsevier, 2008, 496 pp.
25. J. Lehmann, *Carbohydrates: Structure and Biology*, 2nd edn. Thieme, Stuttgart, 1998, 274 pp.
26. D. Horton, Preface, *Carbohydr. Res.*, 341 (2006) 7–8.

AUTHOR INDEX*

A

Aboitiz, N., 164
Achmatowicz, O., 196
Adams, M., 23
Agrawal, P.K., 151
Agusti, R., 171, 191–192
Alexeeva, V.G., 67
Alexeev, Yu.E., 67
Aljahdali, A.Z., 196
Al-Masoudi, N.A.L., 153–154
Anderson, E., 20, 38
Anderson, L., 171
Anet, E.F.L.J., 48
Angyal, S.J., 38, 93, 107, 123, 127, 168, 176, 181, 192, 197–198, 202–203
Antonakis, K., 107
Archibald, A.R., 52
Arnott, S., 1–12, 199
Aronson, N.N. Jr., 67
Arya, D.P., 163–164, 192
Asensio, J.L., 151
Aspinall, G.O., 27, 38, 135, 171–172, 173–174
Atkins, M., 17–18
Augé, C., 127

B

Bacon, J.S.D., 76
Baddiley, J., 52, 198
Baer, H., 59, 171
Bailey, R.W., 44
Baker, B.R., 73, 74
Baker, D.C., 154
Ball, D.H., 56, 59

Ballou, C.E., 27, 80, 101
Banaszek, A., 103
Banks, W., 46
Barker, G.R., 26, 32, 65
Barker, H. A., 80
Barker, R., 40
Barker, S.A., 101
Barnett, J.A., 54, 67, 80, 101
Barnett, J.E.G., 99
Barreto-Bergter, E., 105
Barrett, E.P., 24
Barry, C.P., 26
Barsky, J., 13–14
Barton, D.H.R., 4, 151
Basu, S., 198
Bayer, E.A., 192
Bayne, S., 32
Beélik, A., 32
Bell, D.J., 24, 76, 77
BeMiller, J.N., 9–9, 36, 54, 125, 181
Berman, E., 110
Bhat, K. Venkataraman, 52
Biermann, C.J., 120
Binkley, R.W., 27, 99, 119
Binkley, W.W., 30, 93
Birch, G.G., 46
Bishop, C.T., 48
Blair, M.G., 20, 27, 157
Bleha, T., 123
Bobbitt, J.M., 32
Bock, K., 104, 107, 119, 164, 201
Böeseken, J., 22
Bognár, R., 127, 129–130
Bonner, T.G., 99
Bonner, W.A., 24, 42
Bonten, E., 180

*This list includes names of some contributing coauthors not specifically listed in the text.

Boons, G.-J., 161
Borsani, G., 180
Bourne, E.J., 23, 26, 65, 84, 87, 88, 101
Bourquelot, É., 46, 47
Bouveng, H.O., 40, 78
Brady, R.F. Jr., 65, 101
Bray, H.G., 22, 27
Brennan, A.J., 135
Bresciani, R., 180
Brewer, F., 175
Brimacombe, J.S., 48, 83, 171
Brinson, K., 107
Buchanan, J.G., 147, 176, 198, 203–204
Buchholz, K., 175
Bundle, D.R., 141
Bushway, A.A., 80
Butterworth, R.F., 65

C

Cadenas, R.A., 78
Caesar, G.V., 36
Cai, F., 171
Caldwell, M.L., 23
Calvin, M., 138, 147, 149–150
Cañada, F.J., 151
Canales, A., 164
Cantor, S.M., 23, 24, 26, 50, 205
Capell, L.T., 15–16, 20, 21, 52, 54, 56, 59, 63, 65, 67, 70, 73, 76, 78, 81, 84, 87, 90, 97, 101, 104, 105, 145
Capon, B., 40
Carr, C.J., 20
Carter, R.D., 123
Casu, B., 110, 153, 191
Černý, M., 86, 101, 103, 157–158
Chabre, Y.M., 175–176, 179–180
Chandrasekaran, R., 138, 199
Chatterjee, D., 135
Chen, M., 86
Chizhov, O.S., 52, 72, 167
Chmielewski, M., 164
Churms, S.C., 62
Cipolla, L., 168
Clamp, J.R., 42
Clarke, M.A., 138, 147, 150
Clarke, R.T., 119
Coates, J.H., 119
Compton, J., 21–22

Conchie, J., 34
Corina, D.L., 67
Courtois, J.E., 46, 127, 130
Coxon, B., 67, 72, 99, 119, 170, 184
Crawford, S.A., 96
Crawford, T.C., 96, 99, 131
Crescenzi, V., 171–172, 174
Cretcher, L.H., 56
Crich, D., 171
Crick, F., 4, 199
Crum, J.D., 36
Csuk, R., 119, 124, 170
Czernecki, S., 140

D

Daïs, P., 115–116, 134
Daleo, G.R., 112–113
Dam, T.K., 175
Danishefsky, S.J., 196
Datema, R., 103, 112–113
David, L., 184
David, S., 127, 138
Davies, D.A.L., 40
Dax, K., 83
D'Azzo, A., 180
Dea, I.C.M., 78, 89
Dean, G.R., 23
de Belder, A.N., 50, 65, 86, 101
Defaye, J., 62, 78, 101
Deitz, V.R., 28
Dekker, R.F.H., 80
de Lederkremer, R.M., 120, 131, 158, 160, 167, 171
Dell, A., 115, 167
Delmer, D.P., 105
Demchenko, A.V., 170–171
Deuel, H.J. Jr., 42
Deulofeu, V., 22, 78, 120, 122
Dey, P.M., 89, 96–97, 107, 112–113
Dickey, E.E., 20
Dill, K., 110, 116, 123
Dimler, R.J., 26, 101
Dondoni, A., 196
Dore, W.H., 191
Dorland, L., 119, 201
Doudoroff, M., 23
Dubach, P., 42
Dulaney, S.B., 191

Durette, P.L., 65
Dutcher, J.D., 46, 89
Dutton, G.G.S., 69, 72, 75, 138, 151, 152

E

Eddy, E.D., 56
Edsall, J.T., 13–14
Edye, L.A., 138
Eggleston, G., 138
El Ashry, El-S.H., 184
Elbein, A.D., 75
Elderfield, R.C., 20, 52
El Khadem, H.S., 46, 50, 63, 99, 135, 141, 146–147, 160, 192
Ellis, G.P., 30, 38
El Nemr, A., 184
Espinosa, J.F., 151
Estramareix, B., 138
Evans, T.H., 21, 24
Evans, W.L., 14, 19

F

Fatiadi, A.J., 160
Feather, M.S., 69
Fedoroňko, M., 72
Fernandez-Bolaños, J.G., 115, 153–154
Fernández, G., 146
Ferrier, R.J., 50, 59, 89, 127, 157, 158, 171–172, 198
Fewster, J.A., 32
Finch, P., 138
Finne, J., 99
Fischer, E., 20, 38, 52, 53, 101, 105, 115, 131, 146–147, 157, 158, 164, 171
Fischer, H.O.L., 14, 19, 40, 44, 45
Fitreman, J., 167
Fletcher, H.G. Jr., 21–22, 23, 24, 73, 78, 79
Flowers, H.M., 101
Fordyce, C.R., 20
Foster, A.B., 17–18, 26, 30, 34, 38, 40, 110, 191
Fox, J.J., 38, 54
Franklin, R., 199
Frascaroli, M.I., 188
Fraser-Reid, B.O., 191–192
French, A.D., 191
French, D., 27, 34, 107–108, 109, 119
Freudenberg, K., 52, 105, 157
Frush, H.L., 141, 144

G

Gabius, H.-J., 164
Gallo-Rodriguez, C., 160
García Fernández, J.M., 146
Garcia González, F.G., 32, 50, 115, 118, 176
García, S.I., 188
Garegg, P.J., 137, 160, 170–171, 172, 180, 181, 183
Garg, H.G., 110, 128, 132
Gautheron, C., 127
Gelas, J., 101
Gelpi, M.E., 78
Giese, B., 151
Glänzer, B.I., 119, 124, 170
Glattfeld, J.W.E., 36
Glaudemans, C.P.J., 78
Godshall, M.A., 147
Goepp, R.M. Jr., 14, 16–17, 19, 21, 22
Goldstein, I.J., 52, 89
Gómez-Sanchez, A., 50, 115
Gonzalez, J.C., 188
Goodman, I., 36
Goodman, L., 54, 73
Gorin, P.A.J., 56, 89, 99, 104, 105, 119
Gottfried, J.B., 23
Gottschalk, A., 23, 84, 85
Green, J.W., 21–22, 52, 158
Greenwood, C.T., 26, 32, 46, 54, 56
Grindley, B., 140
Grisebach, H., 89
Gros, E.G., 120
Groves, P., 164
Grynkiewicz, G., 103
Guo, Z., 191–192
Gurin, S., 21–22
Guthrie, R.D., 17, 42, 54, 157–158
Györgydeák, Z., 163

H

Haines, A.H., 83, 101
Hale, K.J., 188
Hall, L.D., 48, 67, 99, 119, 184, 187
Hanessian, S., 52, 65, 83, 167
Hann, R., 46
Hardegger, E., 99, 100
Harding, S.E., 123
Harris, E.E., 22, 69

Harris, J. F., 69
Haskins, J.F., 21
Hassid, W.Z., 23, 46, 65, 80, 82
Haworth, Sir W.N., 14, 19, 23, 24–25, 26, 50, 135, 138, 188–189, 200, 205, 206
Hayes, C.E., 89
Haynes, L.J., 30, 46, 50, 83, 89, 99, 105
Hedgley, E.J., 44
Hehre, E.J., 27, 146–147, 160, 162
Heidelberger, M., 48
Helferich, B., 21–22, 26, 115, 117–118
Hemmer, R., 127
Herp, A., 97
Heuser, E., 40, 41
Hevey, R., 197
Heyns, K., 44
Hibbert, H., 21, 42, 43, 124
Hicks, K.B., 119
Hickson, J.L., 42
Hilton, H.W., 52
Hindert, M., 28
Hirst, Sir Edmund L., 21, 24–25, 80–81, 87, 91
Hoberg, J.O., 157
Hockett, R.C., 21, 58
Hodge, J.E., 30, 141, 142–143
Hoffman, D.O., 21
Honeyman, J., 26, 30, 34
Horton, D., 16–17, 22, 38, 40, 46, 48, 58, 59, 63, 65, 67, 73, 76, 78, 80, 81, 84, 87, 90, 93, 97, 99, 101, 104, 105, 108, 112, 113, 116, 120, 123, 125, 128–129, 132–133, 135–136, 138, 141, 146, 147, 151, 154, 157–158, 161, 164, 168, 172, 176, 181, 185, 189, 192, 194–195, 198, 206
Hough, L., 32, 40, 42
Hounsell, E., 132
How, M.J., 48
Huang, X., 191
Huber, G., 27
Hudson, C.S., 14, 19, 20, 21–22, 23, 24, 26, 28, 29, 42, 44, 46, 50, 58, 78, 131, 135, 171, 200, 205
Huggard, A.J., 30, 110, 191
Hughes, R.C., 176
Hullar, T.L., 52
Hutson, D.H., 46, 80

I

Igarashi, K., 86
Ikehara, M., 92
Inch, T.D., 67
Inokawa, S., 107
Irvine, Sir James C., 26
Isbell, H.S., 56, 59, 107, 135, 136, 141
Islam, T., 168
Ito, Y., 194

J

Jacoby, K., 13–14, 145, 200, 205
Jamieson, G.A., 67
Jansson, P.-E., 158
Järnefelt, J., 99
Jarosz, S., 164, 167–168
Jeanes, A., 93, 141, 143–144
Jeanloz, R.W., 17–18, 24, 32, 110, 132, 172, 176, 177–178
Jeffrey, G.A., 48, 62, 75, 78, 80, 84, 87, 97, 99, 110–112, 151, 154, 155–156, 191
Jennings, H., 105
Jiménez-Barbero, J., 151, 164, 198
Johnson, W.C. Jr., 115
Johnson, W.J., 13–14
Jones, D.M., 21, 48
Jones, J.K.N., 22, 32, 87, 105, 106

K

Kaji, A., 107
Karabinos, J.V., 20, 26, 28, 93, 95
Karban, J., 163
Kärkkäinen, J., 99
Karplus, M., 170
Kashimura, N., 120
Keglević, D., 92
Kendrew, Sir John., 4–5, 13–14
Kennedy, J.F., 73, 101
Kenne, L., 176
Kent, P.W., 21–22
Kertesz, Z.I., 23, 83–84
Khan, R., 83, 101
Khorana, H.G., 92, 198
Kiliani, H., 44
Kinzy, W., 131, 170–171
Kiss, J., 59, 73, 103
Klemer, A., 42, 99
Klenk., 103

Knirel, Y.A., 123, 158, 168
Kochetkov, N.K., 52, 69, 72, 112–113, 164, 167, 168, 169
Koenig, J.L., 112
Komano, T., 120
Kort, M.J., 62–63
Kosma, P., 196
Kowkabany, G.N., 27
Krantz, J.C. Jr., 20
Kroutil, J., 163
Krusius, T., 99
Kuhn, R., 59, 60
Kunz, H., 132
Kuszmann, J., 70

L

La Ferla, B., 168
Laidlaw, R.A., 26
Laland, S., 40
Larm, O., 83, 176, 181
Lederburg, J., 13–14
Lederer, E., 42, 112–113
Lederkremer, R.M. de., 191–192
Lee, C.K., 116
Lee, Y.C., 96, 175–176
Legler, G., 125, 201
Lemieux, R.U., 27, 131, 134, 151, 158, 159, 170, 184
Lespieau, R., 21, 103, 196
Levene, P.A.T., 34, 35
Levi, I., 22, 83
Levvy, G.A., 34, 38, 69, 107
Lewandowski, B., 167
Ley, S.V., 191–192
Lezica, R.P., 112–113
Liggett, R.W., 28
Lincoln, S.F., 119
Lindahl., 153
Lindberg, A., 40, 78, 83, 124, 176
Lindberg, B., 161, 172, 176, 178, 181
Lindner, B., 131–132
Ling, C.C., 197
Lion, S.-C., 112–113
Lipták, A., 127
Li, S.-C., 103
Li, Y.-T., 103, 112–113
Lohmar, R., 22
Lönngren., 72, 78

Lowary, T., 176, 201
Lucas, S.D., 175
Lundt, I., 164
Lyon, N.B., 128

M

MacAllister, R.V., 92
MacMillan., 196
Maeda, K., 125, 154
Maher, G.G., 30
Malaprade, L., 105, 163
Malhotra, O.P., 42, 107
Manley-Harris, M., 137–138
Manners, D.J., 34, 44, 76, 90
Marchessault, R.H., 54, 84, 89–90, 93, 104, 138, 191
Marino, C., 158, 167
Markham, A.F., 92
Marković, O., 83–84
Marradi, M., 179–180
Marshall, J.J., 75, 78, 112, 132
Marshall, R.D., 63
Marsh, C.A., 34, 38, 107
Martín-Lomas, M., 179–180
Matheson, N.K., 112
Mathlouthi, M., 112
Mawhinney, T.P., 180
Maya, I., 153–154
McCarthy, J.F., 54
McCasland, G.E., 50, 124
McCleary, B.V., 112
McCloskey, C.M., 34
McColloch, R.J., 23, 83–84
McDonald, E.J., 21, 137–138
McGale, E.H.F., 59
McGinnis, G.D., 65, 107, 113
Mehltretter, C.L., 27
Mehta, N.C., 42
Mellet, O., 146
Messner, P., 196
Mester, L., 36, 116
Meyer, K.H., 33, 191
Michalski, J.-C., 184
Micheel, F., 42, 99
Millane, R.P., 9–9
Miller, R.E., 24
Mills, J.A., 30, 93, 94–95
Milne, E.A., 56

Mirelman, D., 192
Mischnik, P., 179, 191
Miyake, T., 154
Mizuno, T., 72–73, 103
Momcilovic, D., 179, 191
Monteiro, M., 153, 171–172
Montgomery, J.A., 44
Montgomery, R., 54
Monti, E., 180
Montreuil, J., 96, 99, 103, 104–105, 184, 186–187
Moody, G.J., 48
Morehouse, M.G., 21
Morgan, J.W.W., 34
Mori, T., 27, 185
Morris, E.R., 7
Morrison, A., 78, 89
Mort, A.J., 123
Moses., 147
Mossine, V.V., 180
Moyer, J.D., 38, 67
Muertgeert, J., 42
Myrbäck, K., 21–22

N

Nakagawa, Y., 194
Nakahara, W., 80
Nánási, P., 127
Neely, W.B., 34, 40, 112
Nelson, D.A., 120
Nelson, R.D., 42, 46, 48
Neuberg, C., 22, 36, 37
Neuberger, A., 63, 84, 132
Neufeld, E.F., 46
Neurath, H., 13–14
Newth, F.H., 24, 30, 99
Nickerson, R.F., 23
Nicotra, F., 168, 175–176
Nikaido, H., 65, 69
Nikolaev., 191–192
Nishimura, S.-I., 185
Nord, F.F., 36
Norris., 192

O

O'Doherty, G., 196
Ogawa, S., 191–192
Ogawa, T., 124

Ohtsuka, E., 92
Olson, E.J., 34
Onodera, K., 108, 120, 121–122
Orenstein, N.S., 78
Orgueira, A.A., 146
Ortiz Mellet, C., 146
Oscarson, S., 181
Overend, W.G., 27, 52, 138, 198

P

Pacsu, E., 20
Pałański, M., 123
Palmacci, E.R., 157, 170–171
Parolis, H., 151
Parrish, F., 56, 59
Pauling, L., 13–14
Paulsen, H., 44, 56, 65, 99
Pavia, A.A., 110
Pazur, J.H., 67, 101, 107–108, 140
Peat, S., 19, 20, 21, 23, 58, 62, 64, 78, 101, 157–158
Peczuh, M.W., 188–189
Pedersen, C., 104, 107, 164, 165–166, 201
Pedersen, H., 107, 119
Penadés, S., 179–180
Penglis, A.A.E., 99, 119, 124
Percheron, F., 127
Percival, E.G.V., 24, 26, 30, 31, 141
Pérez, S., 161, 179, 191
Peri, F., 168
Perlin, A.S., 38, 56, 115–116, 163
Pernet, A.G., 83
Perry, M.B., 195
Perutz, M., 4–5
Phillips, G.O., 42, 46, 96, 99, 125
Pickford, R., 167
Pigman, W.W., 13–14, 16–17, 20, 21, 22, 23, 56, 59, 87, 97, 98, 107, 145, 157–158, 171, 205
Plante, O.J., 157, 170–171
Polglase, W.J., 30
Poveda, A., 151
Pozsgay, V., 151
Praly, J.-P., 151
Priddle, J.E., 40
Pridham, J.B., 44, 50
Purves, C.B., 21, 22, 50, 56, 57, 83

Q

Queneau, Y., 167–168

R

Rama Krishna, N., 151
Rauter, A.P., 175
Rauvala, H., 99
Raymond, A.L., 20, 46
Redaelli, C., 168
Rees, D.A., 5–6, 7, 59, 185
Reeves, R.E., 24, 36
Reichstein, T., 44, 52
Reisner, Y., 192
Rendleman, J.A. Jr., 52
Rexova-Benkova, L., 83–84
Reynolds, D.D., 24
Reynolds, J.C., 167
Richards, G.N., 80, 137–138
Richardson, A.C., 188, 190
Richtmyer, N., 20, 23, 24
Rietschel, E.Th., 131–132
Rizzo, R., 171–172
Robertson, J.M., 4
Robyt, J.F., 134–135
Rodrigues, J.A., 167
Roscher, N., 141
Roseman, S., 189, 198, 204
Rosenstein, R.D., 48, 62
Rosenthal, A., 56
Roy, R., 175–176
Rundle, R.E., 107–108

S

Saha, J., 188–189
Sala, L.F., 188
Samain, D., 179, 191
Sánchez, A.G., 50, 176
Sandford, P., 93, 141
Sands, L., 20
Santos, M., 175
Sarko, A., 54, 84, 138
Sattler, L., 21–22
Schäffer., 160, 196
Schauer, R., 103, 158, 180, 201
Schilling, C., 140–141, 160
Schmidt, R.R., 131, 137, 170–171, 191–192
Schoch, T.J., 14, 20, 42
Schuerch, C., 101
Schwarz, R.J., 103
Schwarz, R.T., 103, 112–113
Scott, W.E., 5–6
Sebastian, S., 179–180
Seeberger, P.H., 157, 158, 170–171, 191–192, 196
Seibel., 175
Serpersu, E., 192
Shafizadeh, F., 32, 36, 56, 65, 78, 107, 108, 113, 114
Sharon, N., 176, 185, 189, 192, 193
Shashkov, A.S., 158
Shi, P., 196
Shibaev, V.N., 69, 112–113
Siddiqui, I.R., 62–63, 141
Sidebotham, R.L., 75
Singh, P.P., 80
Smirnova, G.P., 112–113
Smith, F., 21, 22, 50, 54, 55, 96, 163
Smoot, J.T., 170–171
Snaith, S.M., 69, 107
Soltzberg, S., 62, 127
Somsák, L., 127
Sowden, J.C., 24, 34, 44, 46, 50, 51
Speck, J.C. Jr., 36
Spedding, H., 48, 112
Spencer, J.F.T., 56, 89
Spohr, U., 131
Sponsler., 191
Sprinson, D.B., 40
Stacey, M., 21–22, 26, 27, 48, 83, 90, 105, 124, 135, 138, 139, 198
Staněk, J. Jr., 86
Stetter, H., 115
Stevens, J.D., 188–189, 197–198
Stoloff, L., 36
Stoss, P., 127
Stowell, C.P., 96, 175–176
Strahs, G., 62, 110–111
Stütz, A.E., 189
Suami, T., 124
Sugihara, J.M., 27, 83, 101
Sumpton, D.P., 167
Sundaralingam, M., 75, 78, 80, 84, 87, 97, 99, 111–112
Sundararajan, P.R., 84, 89–90, 93, 104, 138, 179, 191

Svensson, S., 72, 78
Swarts, B.M., 191–192
Szarek, W.A., 69, 105, 151
Sztaricskai, F., 127

T

Talley, E.A., 24
Taravel, F.R., 134, 170
Taylor, A.M., 167
Teague, R.S., 27
Tettamanti, G., 180
Theander, O., 44, 120, 158
Theobald, R.S., 40
Thomas, H.J., 44
Thomas-Oates, J., 167
Thompson, A., 48, 49
Timell, T.E., 48, 50
Tipson, R.S., 17–18, 20, 27, 28, 32–33, 34–35, 36–37, 38–39, 40–41, 42–43, 44–45, 46, 48, 50, 52, 54, 56, 58, 59, 63, 65, 67, 70, 73, 76, 78, 81, 84, 87, 90, 93, 97, 99, 101, 104, 105, 108, 112, 113, 116, 120, 123, 125, 128, 132, 133, 145, 201, 206
Todd, A.R. (Lord)., 138, 147, 148–149
Todt, K., 56
Tokuzen, R., 80
Tomasík, P., 123, 135, 140–141, 160, 194–195
Tsuchiya, T., 124, 125
Tsvetkov, Y.E., 158
Turvey, J.R., 50, 62
Tvaroška, I., 123, 134, 170
Tvetskov, Yu.E., 158

U

Ueda, T., 54
Umezawa, H., 75, 125, 126
Umezawa, S., 75, 151, 154, 156, 192
Unger, F.M., 99, 153
Usov, A.I., 185

V

Vaino, A.R., 151
van Boom, J., 160–161
Van Doorslaer, S., 188
van Halbeek, H., 119, 201
Varela, O., 131, 146, 158
Vargha, L., 70, 71
Venerando, B., 180
Verstraeten, L.M.J., 54, 137–138
Vinogradov, E.V., 123
Vishwakarma, R.A., 191–192
Vliegenthart, J.F.G., 104–105, 119, 201
Von Den Bruch, K., 132
von Itzstein, M., 168
von Sonntag, C., 96

W

Wallenfels, K., 42, 107
Wander, J.D., 80
Watson, R.R., 4, 78, 199
Webber, J.M., 17–18, 40, 44, 185
Weidmann, H., 83
Weigel, H., 46, 87
Weiss, A.H., 72–73
Weiss, E., 52
Wellington, N.Z., 89
Wempen, I., 38
Whistler, R.L., 16–17, 18, 20, 23, 34, 36, 42, 80, 86, 161, 180, 182–183
Whitehouse, M.W., 36, 103
Whitfield, D.M., 170
Wiejak, S., 123
Wiggins, L.F., 22, 23, 108, 127
Wightman, R., 198
Wilkie, K.C.B., 93
Wilkins, M.H.F., 4, 199
Williams, J.M., 78
Williams, N.R., 62, 101
Willis, B., 163–164, 192
Wise, L.E., 40
Witczak, Z.J., 112, 140
Wolfrom, M.L., 13–14, 15, 16–17, 18, 20, 21, 22, 23, 24, 26, 27, 28–29, 30–31, 32, 34, 36, 38, 40, 42, 44, 46, 48, 50, 52, 54, 56, 58, 62, 65, 66, 132, 135, 137, 145, 171, 200, 201, 205, 206
Wrodnigg, T., 189
Wu, B., 171

X

Xavier, N.M., 175

Y

Yamamoto, H., 107
Yoshimura, J., 107

Z

Zähringer, U., 131–132, 158
Zamojski, A., 103, 160–161, 164, 166, 196
Zaranyika, M.F., 135
Zemplén, G., 38, 39
Zhdanov, Yu.A., 67
Zhong, Y., 196
Zilliken, F., 36, 103
Zoltan., 163
Zorbach, W.W., 52, 67, 68

SUBJECT INDEX

Advances in Carbohydrate Chemistry
  articles (*see* Articles)
  background
    carbohydrate nomenclature and
      indexing, 14–16
    carbohydrate reference books, 16–17
    launching of, 13–14
    setting up research journal, 17–18
Arnott, Struther
  carrageenan-like double-helical structures, 6
  glycosaminoglycans of vertebrate
    connective tissues, 6–7
  investigation on limonin ($C_{26}H_{30}O_8$), 4
  molecular architectures, allomorphs, 7
  Watson–Crick base-pairing scheme, 5
  X-ray diffraction, 4
Articles
  acetals and ketals of the tetritols, pentitols,
    and hexitols, 26
  acetolysis, process of, 54
  acid-catalyzed hydrolysis of glycosides, 54
  acoritic acid as a byproduct of sugar
    manufacture, 24
  action of alpha amylases, 23
  acylated nitriles of aldonic acids and their
    degradation, 22
  acyl esters of carbohydrates, 78
  acyl esters, reactions with ammonia, 78
  affinity chromatography, 101
  alditol anhydrides, 62
  alditol metabolism, 20
  aldonolactones as chiral synthons, 131
  2-(aldopolyhydroxyalkyl)benzimidazoles,
    24
  algal polysaccharides, 185
  aliphatic alcohols and starch, 140–141

alkali metals and alkaline-earth metals with
  carbohydrates, 52
alpha amylases, action of, 23
altrose, 20
Amadori rearrangement, 180
2-amino-2-deoxy sugars, 26
aminoglycoside antibiotics, mechanism of
  resistance to, 75, 163–164
aminoglycoside-modifying enzymes
  (AGMEs), 192
amino sugars, reaction with β-dicarbonyl
  compounds, 50
amino sugars, 38, 46
amino sugars, aspects of the chemistry
  of, 40
amino sugars in antibiotics, 46
amino sugars, reaction of, 176
amino sugars, tabulated properties of, 40
amylases, alpha and beta, 195
amylose, commercial production of, 42
analogues of cyclic sugars, 124
analysis of polysaccharide structure, 112
Angyal, Stephen J., portrait, 199
anhydrides of alditols, 62
anhydrides of pentitols and hexitols, 23
1,6-anhydro derivatives of aldohexoses, 86
1,6-anhydrohexofuranoses, 26
1,6-anhydrohexoses, 86
2,5-anhydro rings, 62
anhydro sugars, 21
anhydrous hydrogen fluoride, 164
anomeric and exo-anomeric effects in
  carbohydrate chemistry, 123
antibiotics, biosynthesis of, 89
antibiotics, broad-spectrum, 151
antibiotics, development of, 125

Articles (*Continued*)
  antibiotic substances, sugar components of, 89
  anti-carbohydrate antibodies, 140
  anticoagulant therapy, 110
  antitumor polysaccharides, 80
  apiose, 78
  apiose and glycosides of the parsley plant, 22
  application of cationic polymerization, 101
  applications of monosaccharide thiocyanates and isothiocyanates, 112
  applications of X-ray crystallography, 134
  aqueous acidic hydrolysis and cleavages of glycosidic linkages, 120
  aqueous degradation of carbohydrates, 120
  L-arabinosidases, 107
  Arixstra®, 191
  Arnott, Struther, portrait, 203
  aromatic amino acids, biosynthesis of, 40
  L-ascorbic acid, 96
  ascorbic acid analogues, 21
  aspects of the chemistry of the amino sugars, 40
  Aspinall, Gerald O., portrait, 173
  aziridines of sugars, 163
  bacteria, carbohydrate constituents of, 40
  bacterial capsular polysaccharides, 105
  bacterial cell-envelope, 196
  bacterial lipopolysaccharides, 131–132
  bacterial polysaccharide chains, biosynthesis of, 112–113
  bacterial polysaccharides, 21, 124, 176, 181
  Baker, Bernard Randall, portrait, 74
  Bell, David J., portrait, 77
  benzyl ethers, 34
  biochemical reductions at the expense of sugars, 22
  biochemistry of plant galactomannans, 89
  biogenesis of cellulose, 65
  biogenesis of monosaccharides, 32
  biomass, utilization of, 120
  biosynthesis and catabolism of glycosphingolipids, 103
  biosynthesis of bacterial polysaccharide chains, 112–113
  biosynthesis of cellulose fiber, 105
  biosynthesis of polysaccharides, 80
  biosynthesis of saccharides, 46, 65
  biosynthesis of the monosaccharides, 32
  blood-group polysaccharides, 22
  Bognár, Reszö, portrait, 129
  boric acid for the determination of the configuration of carbohydrates, use of, 22
  boronates of carbohydrates, 89
  Bourne, Edward J., portrait, 88
  Bourquelot, Émile, portrait, 47
  branched-chain sugars, 107
  broad-spectrum antibiotics, 151
  bromination reactions of carbohydrates, 127
  Buchanan, J. Grant, portrait, 200
  Calvin, Melvin, portrait, 149
  cancer chemotherapy, 80
  cane juice and cane-final molasses, 27
  capsular polysaccharides, bacterial, 105
  carba sugars, 124
  carbohydrate and fat metabolism, 21
  carbohydrate and thiocarbonate esters, 40
  carbohydrate-based nanoparticles, 179–180
  carbohydrate-binding properties of lectins, 194
  carbohydrate chemistry, crystal-structure analysis in, 110–111
  carbohydrate components of bacteria, 40
  carbohydrate–lectin interaction, 197
  carbohydrate nanotechnology, 179–180
  carbohydrate nomenclature, 137
  carbohydrate nomenclature, current usage, 16
  carbohydrate orthoesters, 20
  carbohydrate polymers, 52
  carbohydrate–protein conjugates, biological recognition, 175
  carbohydrate–protein interactions, plant and animal species, 89, 92
  Carbohydrate Research, 204
  carbohydrate-separation techniques, 34
  carbohydrate sequences of bacterial polysaccharides, 181
  carbohydrates, nomenclature of, 58, 145–146

carbohydrates, nucleosides, and nucleotides, crystal structures of, 87, 99, 111–112
carbohydrates of the cardiac glycosides, 20
carbohydrates, structural investigation of, 105
carbohydrate structural methodology by C-13 NMR, 110
carbohydrate sulfonates, 56
carbohydrate technology, 135
carbon-13 NMR spectroscopy, 116
carbon-13 nuclear magnetic relaxation, 134
carbon–proton coupling constants, 134
carbon-13 techniques, 105
cardenolides, 52
cardiac glycosides, 44
cardiac glycosides, carbohydrates of, 20
cardiac glycosides, sugars in, 44
catabolism of glycosphingolipids, biosynthesis and, 103
cationic polymerization, application of, 101
Chemical Abstracts, 15–16
cell adhesion, 197
cell-envelope, bacterial, 196
cellobiose derivatives, three-dimensional structures, 191
cell-surface lectins, 197
cellulose and plant cell-walls, 65
cellulose and starch, chemical derivatization of, 179
cellulose and starch, chemical modifications of, 179
cellulose, constitution of, 21–22
cellulose, crystallinity of, 23
cellulose esters, 20
cellulose ethers, 21
cellulose fiber, biosynthesis of, 105
cellulose macromolecule, 48
cellulose, structure, morphology and solubilization, 179
cellulose, structure of, 191
characterization of carbohydrates, 119
chemical degradation of polysaccharides, 78
chemical derivatization of cellulose and starch, 179
chemically reactive derivatives of polysaccharides, 73

chemical structure of heparin, 110
chemistry and interactions of seed galactomannans, 78
chemistry of carbohydrates Wittig reaction in, 67
chemistry of maltose, 101
chemistry of nucleosides and nucleotides, 92
chemistry of ribose, 24
chemistry of streptomycin, 21–22
chemistry of sugar aziridines, 163
chemoselective neoglycosylation, 168
chitin, 40
cholinergic agents from carbohydrates, 67
cholinergic drugs, 67
chromatography, 27
circular dichroism, 115
Clarke, Margaret A., portrait, 150
^{13}C-NMR of monosaccharides, 205
^{13}C-NMR spectroscopy of monosaccharides, 107, 119
^{13}C-nuclear magnetic resonance spectroscopy of polysaccharides, 99
column chromatography of sugars, 30
commercial production of dextrose, 23
complexes between carbohydrates and alkali and alkaline-earth metal ions, 52
complexes of starch, 140–141
complexes of the cyclomalto-oligosaccharides, 119
composition of cane juice and cane final molasses, 22
condensation of phenylboronic acid, 89
conformational analysis of sugars, 30, 65
conformation of C-glycosyl compounds, 151
conjugated oligosaccharide determinants, 157
constituents of cane molasses, 21–22
constitution of cellulose, 21–22
construction of glycosidic linkages, 175
Courtois, Jean Émile, portrait, 130
Crescenzi, Vittorio, portrait, 174
crystalline glycosidase, 42
crystallinity of cellulose, 23
crystallographic method, 75, 78
crystallographic structure-determination, 191

Articles (*Continued*)
  crystallographic-structure studies, 111
  crystallography of sugars, 110–111
  crystal-structure analysis, 48, 62
  crystal-structure analysis in carbohydrate chemistry, 110–111
  crystal structure of polysaccharides, 104
  crystal structures, 75, 78, 80, 87, 97
  crystal structures of carbohydrates, nucleosides, and nucleotides, 87, 111–112
  cumulative indexes, 145
  cuprammonium glycoside complexes, 24
  cyclic acetals, 50
  cyclic acetals of aldoses and aldosides, 86
  cyclic acetals of ketoses, 65
  cyclic acetals, particularly 1,3-dioxolanes and 1,3-dioxanes, 101
  cyclic acyloxonium ions, 65
  cyclitols, chemistry of, 21–22, 197–198
  cyclitols, 21–22
  cyclodextrin inclusion complexes, 119
  daumone, 196
  deamination of amino sugars, 78
  deamination of carbohydrate amines and related compounds, 78
  degradation to dicarbonyl compounds, 48
  dehydration of sugars, 69
  dehydration reactions of sugars, 69
  deoxyhalogeno sugars, 69
  deoxyinositols, 50
  deoxy sugars, 52, 167
  deoxy sugars, tabulated properties of, 65
  depolymerization by anhydrous hydrogen fluoride, 123
  2-desoxy sugars, 22, 27
  desulfurization, 23
  desulfurization by Raney nickel, 23
  Deulofeu, Venancio, portrait, 122
  development of antibiotics, 125
  development of carbohydrate chemistry, 120
  development of dextran and xanthan, 141
  dextran and xanthan, 141
  dextrans, 40, 75
  "dextrins", manufacture, 123
  "dextrins", meaning, 195

dextrose, commercial production of, 23
dialdehydes ring structure, 42
dianhydrides of D-fructose, 137–138
1,4:3,6-dianhydrohexitols, 127
dicarbonyl sugars, 44
Dictionary of Carbohydrates, 17
Diels–Alder cycloaddition, 164
difructose anhydrides, 21
disaccharide α,α-trehalose, 75
dithioacetals of sugars, 80
Dutton, Guy G.S., portrait, 152
effects of plant-growth substances on carbohydrate systems, 52
electrochemistry of carbohydrates, 72
electrophoresis, 46
element fluorine, 124
β-eliminative degradation of carbohydrates containing uronic acids, 73
empirical variants of the Karplus equation, 170
enzymatic degradation of starch and glycogen, 21–22, 44
enzymatic methods for structure determination, 75
enzymatic specificity, principles underlying, 23
enzymatic synthesis of starch and glycogen, 21–22
enzymatic synthesis of sucrose, 23
enzymatic transformations, 194–195
enzyme and substrate engineering, 175
enzymes acting on carbohydrates, 107
enzymes acting on pectic substances, 23
enzymes as synthetic tools, 127
enzyme structure, 67
enzymatic degradation, 78
extracellular polysaccharides of microorganisms, 93
family of glycoproteins, 116
fast-atom bombardment, 115, 167
fiber X-ray crystallography, 179
Fischer cyanohydrin synthesis, 20, 131, 171
Fischer, Emil, portrait, 53
Fischer, Hermann O. L., portrait, 45
Fischer stereo-formulas, history of, 21–22
Fletcher, Hewitt G., portrait, 79

fluorinated carbohydrates, 99
fluorine, applications in carbohydrates, 42
fluorine atypical halogen, 124
fluorine chemistry, 42
formation of furan compounds from hexoses, 24
formation of polysaccharide gels and networks, 59
formazan reaction on sugar hydrazone structures, 36
formose reaction, 72–73
four-carbon saccharinic acids, 36
Fourier-transform algorithm, 112
fractionation of starch, 20, 42
French, Dexter, portrait, 109
Friedel–Crafts and Grignard reactions in carbohydrates, 24
fructose, 54
fructose and its derivatives, 26
fructose and dianhydrides, 54, 137–138
D-fructose metabolism, 86
Frush, Harriet L., portrait, 144
fucose enantiomers, 101
functionalized polysaccharides, 73
fungal polysaccharides, structure of, 56
furan compounds from hexoses, formation of, 24
galactomannans of plants, 89
D-galactose, methyl ethers of, 24
Garegg, Per Johan, portrait, 183
gas–liquid chromatography, 48
gas–liquid chromatography of sugars, 69, 75, 103
gels and networks of polysaccharides, 59
"Geneva"names, 14–15
glucansucrase enzymes, 134–135
D-glucofuranurono-6,3-lactone, reactions of, 83
glucosamine hydrobromide, structure of, 154
D-glucose and malto-oligosaccharides, 92, 112
D-glucose—D-fructose isomerization by immobilized enzymes, 92
D-glucuronic acid, 27
D-glucuronic acid and its metabolism, 27
D-glucuronic acid, animal conjugates of, 27

D-glucuronic acid, glycosides of, 92
D-glucuronic acid in metabolism, 22
β-glucuronidase, 38
glycals, 26
glycobiology of *Trypanosoma cruzi*, 171
"glycoblotting" technique, 185
glycoclusters and glycodendrimers, 175–176
glycoconjugates of the cell envelope, 196
glycofuranosides, 52
glycogens, 34
glycol cleavage by periodate, 42
glycol-cleavage reaction, 163
glycolipids, nomenclature of, 147
glycolipids of acid-fast bacteria, 42
glycolipids of marine invertebrates, 112–113
glycol-splitting reagents, 42
glycomics and glycobiology, 185
glycopeptides, structure of, 110
glycophorins, carbon-13 NMR of, 116
glycoprotein enzymes, 67
glycoprotein glycans, 96
glycoproteins, 96
glycoproteins, structure and metabolism of, 63
glycopyranosyl esters of nucleoside pyrophosphates, 65
glycopyranosyl oxacarbenium ions, 170
glycosaminoglycan heparin, 110, 191
glycoside cluster effect, 175–176
glycoside hydrolase mechanism, 125
glycosides, 34
glycosides, acid-catalyzed hydrolysis of, 54
glycosides of D-glucuronic acid, 92
glycosides, simple, 34
glycosidic coupling methodology, 137
glycosphingolipids, 59
glycosylamines and their rearrangement, 30
glycosylation, 196
C-glycosyl, 196
glycosyl azides, 163
C-glycosyl compounds, 83
C-glycosyl compounds in plants, 50
glycosyl esters of nucleoside pyrophosphates, 80

Articles (*Continued*)
glycosyl halides, 30
C-glycosyl natural products, 46
glycosylphosphatidylinositol (GPI), 191–192
glycosyl ureides, 36
González, Francisco García, portrait, 118
Gottschalk, Alfred, portrait, 85
GPI anchors, 191–192
granular adsorbents for sugar refining, 24
gums and mucilages, structure of, 59
halogenated carbohydrates derivatives, 54
Hall, Laurance David, portrait, 187
Hardegger, Emil, portrait, 100
Hassid, Zev, portrait, 82
Haworth, Walter, Sir Norman, portrait, 25
Hehre, Edward J., portrait, 162
Helferich, Burckhardt, portrait, 117
helical biopolymer, 107–108
hemicellulases, 80
hemicelluloses, 38, 48, 80, 93
hemicelluloses in wood, 48
hemicelluloses of Gramineae, 93
heparin, 30, 110, 153
heparin, pentasaccharide subunits, 191
heparin, structure and biological activity of, 153
Heuser, Emil, portrait, 41
hexitols and some of their derivatives, 22
hexitol synthesis, 21
Hib vaccine, 168
Hibbert, Harold, portrait, 43
hidden resonance problem, 72
higher-carbon sugars, 44
high-performance liquid chromatography, 48
high-pressure liquid chromatography, 119
high-temperature transformation of carbohydrates, 120
Hirst, Sir Edmund L., portrait, 91
^{1}H-NMR glycoprotein, 205
Hodge, John E., portrait, 142
homoglycan synthesis, 171
honey, sugars of, 62–63
Hudson, Claude Silbert, portrait, 29
human urine, protein–carbohydrate conjugates in, 59
hyaluronic acid, 34

hydrazine derivatives of sugars, 116, 146–147
hydrazones and osazones, 21–22
hydrogen-bonding interactions of carbohydrates, 112
hydrogen-isotope-labeled sugars, 67
hydrogen peroxide, reactions with carbohydrates, 48
hydrolysis of glycosides, by acid and enzymes, 125
hydrolytic and isomerizing enzymes, 92
hydrophilic and hydrophobic interactions, 164
hydroxyglycals, 27
hydroxyl groups, relative reactivities of, 83
in vitro polymerization of formaldehyde, 72–73
imino sugars, 189, 196
immunochemistry, 48
immunoglobulin–polysaccharide interaction, 78
impact factor, 205
indolizidine alkaloids, 196
industrial aspects of carbohydrates, 28
infections by gram-negative pathogens, 131–132
infrared spectroscopy, 34, 48
inhibitors of sialidase, 168
inorganic cations, 188
inositols, 38, 160
International Carbohydrate Symposia, 168
International Sugar Research Foundation, 83
iodine in carbohydrate transformations, 151
ionizing radiation, action on carbohydrates, 42
ionizing radiation, behavior of carbohydrates, 96
irradiation of starch, 135
Isbell, Horace Smith, portrait, 136
isotopes of hydrogen, 67
isotopic tracers in carbohydrate metabolism, 21–22
Jeanes, Allene, portrait, 143
Jeanloz, Roger W., portrait, 177
Jeffrey, George A., portrait, 155

Jones, John Kenyon Netherton, portrait, 106
Karabinos, Joseph V., portrait, 95
Ketonic ester, 32
β-ketonicester–monosaccharide condensation products, 32
ketonucleosides, 107
ketoses, cyclic acetals of, 65
Kochetkov, Nikolai K., portrait, 169
Koenigs–Knorr reaction, 86
Koenigs–Knorr synthesis, 86, 131
kojic acid, 32
Kuhn, Richard, portrait, 60
lactose, 42
lead tetraacetate oxidation, 38
lectin mimics, 194
lectins, 89
Lemieux, Raymond Urgel, portrait, 159
Levene, Phoebus Aaron Theodore, portrait, 35
Lille Glycobiology School, 184
Link, Karl Paul (G.), portrait, 102
Lindberg, Bengt, portrait, 178
Linked-Atom Least-Squares, 202
lipid A, 131–132
lipo-oligosaccharide antigens, 135
lipopolysaccharides, bacterial, 131–132
lipopolysaccharides of *Helicobacter pylori*, 153
lipopolysacharides of O-antigens, 195
"lock and key" concept, 131
macromolecules, 59
macrostructure of mucus glycoproteins, 123
magnetic resonance imaging (MRI), 184
Maillard (nonenzymic) browning reaction, 38
Maillard reaction, 38, 180
Malaprade's periodate cleavage in structural investigation of carbohydrates, 105
maltose, chemistry of, 101
mammalian proteoglycans, 151
mannans and rhamnans, total synthesis of, 171
D-mannose, methyl ethers of, 27
α-D-mannosidase, 69

D-mannuronic acid, 56
mass spectrometry in structural analysis of natural carbohydrates, 72
mass spectrometry of carbohydrates, 52
measurement of proton spin–lattice relaxation rates, 115–116
mechanisms of acid hydrolysis, 125
mechanisms of reaction of sugars, 135
melezitose, 27
melezitose and turanose, 21
metabolism and functions of sialic acids, 103
metabolism of carbohydrates and fat, 21
metabolism of D-fructose, 86
metal ion–carbohydrate interaction, 123
methodology for glycosidic coupling, 157
Methods in Carbohydrate Chemistry, 16–17
methylation analysis and periodate oxidation, 104
methyl ethers, 27, 30
methyl ethers of aldopentoses and of rhamnose and fucose, 26
methyl ethers of D-galactose, 24
methyl ethers of D-glucose, 23
methyl ethers of D-mannose, 22
methyl ethers of sugars, 30
Meyer, Kurt H., portrait, 33
microscale analysis of carbohydrate structures, 176
Mills, John A., portrait, 94
DL mixtures, 101
modern reaction mechanisms, 48
molecular recognition, 140
molecular-sieve techniques, separation of macromolecules by, 62
mono-and oligo-saccharides and polysaccharide technology, 160
mononucleotides, 54
monosaccharide components, utilization of, 101
monosaccharides and oligosaccharides, 196
monosaccharides, biosynthesis of, 32
monosaccharides, *de novo* synthesis, 196
monosaccharides, reactions of, 115
monosaccharide structures, 189

Articles (*Continued*)
  monosaccharide sugars, 163
  monosaccharide thiocyanates and isothiocyanates, applications of, 112
  "mucopolysaccharides", 21
  Montreuil, Jean, portrait, 186
  murine myeloma immunoglobulins, interaction of homogeneous, 78
  mutarotation of sugars, 56, 59
  mutarotation of sugars in solution, 59
  mycobacterial polysaccharides, 21–22
  mycobacteria, surface glycolipids of, 135
  natural cryptands, unique group of, 119
  nature of the plant cell-wall, 107
  neighboring-group participation in sugars, 54
  neoglycoconjugates, 175–176
  neoglycoproteins, 96, 146
  Neuberg, Carl, portrait, 37
  neuraminic acid, structure of, 198
  neuraminulosonic acid, 103
  nitrates of sugar, 34
  nitriles, acylated, of aldonic acids, 22
  nitrogen heterocycles from sugars, 63
  nitrogen or sulfur in the sugar ring, 56
  nitromethane and 2-nitroethanol in syntheses, 24
  nitro sugars, 59
  NMR spectroscopy, 134
  nomenclature of carbohydrates, 58, 145–146
  nomenclature of carbohydrates, 1953 rules, 26–27, 28
  nomenclature of glycolipids, 147
  non-aqueous solvents for carbohydrates, 67
  non-aqueous solvents for sugars, 67
  noncytotoxic polysaccharides, 80
  nonenzymatic browning, 141
  nonulosaminic acids, 36
  nonulosonic acids, 158
  nuclear magnetic resonance, 48
  nuclear magnetic resonance of sugars, 67
  nuclear magnetic resonance spectroscopy, 48
  nuclear magnetic resonance spectroscopy of fluorinated monosaccharides, 119
  nucleic acids, 20, 32
  nucleosides, 44
  C-nucleosides, 83
  nucleosides and nucleotides, 44
  nucleosides and nucleotides, chemistry of, 92
  nucleosides and nucleotides, crystal structures of, 75, 97
  nucleotides, 44
  nucleotides and nucleosides of sugars, 75, 97
  nucleotide sugars, 69
  obituary reports, 205
  olefin chemistry, 50
  oligo-and poly-saccharides, 120
  oligosaccharide determinants of glycoproteins, 132
  oligosaccharides, 27, 107, 119, 175
  oligosaccharides, newly isolated, 44
  oligosaccharide synthesis, 24
  O-linked glycopeptides, 132
  Onodera, Konoshin, portrait, 121
  organic molecules/ions, crystallographic methods, 80
  organotin compounds, 140
  oriented polysaccharide fibers, X-ray analysis, 138
  orthoesters of carbohydrates, 20
  "osazones", sugar, 50
  osazones, sugar hydrazones and, 21–22
  osones, 32
  osotriazoles, 46
  oxidation of carbohydrates, 158
  oxidation of sugars by halogen, 21–22
  oxidation procedure, 44
  oxidative and degradative reactions of sugars and polysaccharides, 158
  oxiranes of sugars, 62
  oxo reaction to carbohydrates, application of, 56
  oxygen/nitrogen atom of amino acids and proteins, 110
  paper chromatography of sugars, 27
  papulacandins, 196
  Peat, Stanley, portrait, 64
  pectic enzymes, 23, 83–84
  pectic materials, 21
  Pedersen, Christian, portrait, 165

Percival, Edward George Vincent, portrait, 31
periodate cleavage, 32
phenomenon of sugar mutarotation, 56
phenylboronic acid, condensation of, 89
phenylhydrazine, sugars in reaction with, 146–147
phosphorus as the ring atom, 107
photochemical reactions of carbohydrates, 99, 119
photochemistry, 46
photosensitive protecting groups, 119
physical methodology for study of sugars, 72
physicochemical methods, 132
Pigman, W. Ward, portrait, 98
plant-growth substances, 52
plant gums and mucilages, 22
plant phenolics, 50
plant "polyuruonides", 20
pneumococcal polysaccharides, 83
polyamides from sugars, 146
polyfunctional nature of sugars, 63
polymerization of anhydro sugars, 101
polynucleotide synthesis, 92
polysaccharide antigens, 78
polysaccharide chains, 138
polysaccharide chemistry, 30
polysaccharide crystal structures, 75, 78, 84, 93
polysaccharide gels and networks, formation of, 59
polysaccharide hydrophilic colloids, 36
polysaccharide–protein binding interactions, 153
polysaccharides, 93, 101, 119
polysaccharides and glycosides, sugars, 145
polysaccharides and oligosaccharides, 96–97
polysaccharides, bacterial, 21, 124
polysaccharides, biosynthesis of, 80
polysaccharides, chemical degradation of, 78
polysaccharides, chemically reactive derivatives of, 73
polysaccharides, chemical synthesis of, 52

polysaccharides, crystal structures of, 104
polysaccharides in solution, physical properties, 46
polysaccharides, molecular size and shape, 26
polysaccharides of fungi and lichens, 105
polysaccharides of seaweeds, 27
polysaccharides, structure and applications of, 181
polysaccharide structural analysis, 112, 123
polysaccharide structure and properties, 40
polysaccharide structure-determination, techniques for, 83
polysaccharides, X-ray crystallographic methods, structures of, 89–90
polysaccharide, X-ray structure of, 54
preparation of neoglycoproteins, 146
principles of molecular-sieve methodology and application in the carbohydrate field, 62
principles underlying enzyme specificity, 23
process of acetolysis, 54
properties of deoxy sugars and their simple derivatives, 65
protecting groups, 101
protective-group strategy and glycosidic coupling, 115
protein–carbohydrate compounds in human urine, 59
protein–carbohydrate interaction, 164
proton magnetic resonance spectroscopy of carbohydrates, 67
proton-NMR spectroscopy, 104–105
proton spin–lattice relaxation rates, measurement of, 115–116
"pseudo-sugar" terminology, 124
psicose, sorbose and tagatose, 26
Purves, Clifford B., portrait, 57
pyranose and furanose ring forms, 188–189
pyrimidine nucleosides, 38
pyrolysis and combustion of cellulosic materials, 56
pyrolysis of cellulose, 56
radicals at the anomeric center, 151
raffinose family of oligosaccharides, 27
Raney nickel, 23

Articles (*Continued*)
  reaction mechanisms in carbohydrates, 27
  reaction mechanisms, modern, 48
  reaction of amino sugars, 176
  reactions of D-glucofuranurono-6,3-lactone, 83
  reactions of monosaccharides, 115
  reactions of sugars, 62–63
  reactivities of hydroxyl groups in carbohydrates, 83
red algae, 185
relative reactivity of hydroxyl groups on carbohydrates, 27
ribose, chemistry of, 24
Richardson, Anthony (Dick) C., portrait, 190
ring-forming reactions of aldoses, 62
ring oxygen atom in sugars, 153–154
Roseman, Saul, portrait, 201
saccharinic acid formation, 34
Schardinger dextrins, 34, 119
SciFinder, 205
seaweed polysaccharides, 22
seed galactomannans, chemistry and interactions of, 78
selectin–carbohydrate interaction, 153
selectins, 153
selenium in synthesis, 140
separation and characterization of carbohydrates, methods of, 69
separation of macromolecules by molecular-sieve techniques, 62
septanose ring forms, 188–189
Shafizadeh, Fred, portrait, 114
Sharon, Nathan, portrait, 193
sialic acid-containing compounds, 180
sialic acids, 103
sialic (nonulosaminic) acids, 36
sialidases, 180
size and shape of polysaccharide molecules, 26
S-layer glycoproteins, 197
"small proteoglycans", 128
Smith, Fred, portrait, 55
"soft" ionization technique, 115
soil, carbohydrates in, 42
solid-phase synthesizer, 157
solid-state synthesis technique, 170–171
Sowden, John C., portrait, 51
spectrometers and techniques of data manipulation, 127
Splenda® (trichloro-*galacto*-sucrose), 188
Stacey, Maurice, portrait, 139
starch, aspects of the physical chemistry of, 32
starch, complexes with inorganic and organic guests, 140–141
starch-degrading enzymes, 195
starch, enzymatic conversion by hydrolases, 194–195
starch enzymes, 56
starch esters, 20
starch, fractionation of, 20
starch hydrolases, 44
starch, isolation from cereal grains, 194
starch nitrate as explosive, 36
starch, thermal degradation of, 54
stereochemistry of monosaccharides, 115
streptomycin, chemistry of, 21–22
structural elucidation of carbohydrates, 115–116
structure and applications of polysaccharides, 181
structure and biological activity of heparin, 153
structure and configuration of sucrose (α-D-glucopyranosyl-β-D-fructofuranoside), 22
structure and metabolism of glycoproteins, 63
structure of cellulose, 191
structure of fungal polysaccharides, 56
structure of glucosamine hydrobromide, 154
structure of glycopeptides, 110
structure of gums and mucilages, 59
structure of neuraminic acid, 198
structures of mammalian proteoglycans, 151
structures of polysaccharides, X-ray crystallographic methods, 89–90
substituted-sugar structure of melezitose, 22
sucrose as precursor, 167–168
sucrose, chemistry of, 83
sucrose, enzymatic synthesis of, 23

sucrose, structure and configuration of, 22
sucrose (α-D-glucopyranosyl-β-D-fructofuranoside), structure and configuration of, 22
sucrose, utilization of, 22
sugar–alkali reactivity, 36
sugar–amine reaction, 62–63
sugar–amino acid linkages, 110, 132
sugar analogues having phosphorus in the ring, 107
sugar aziridines, chemistry of, 163
sugar carbonate and thiocarbonate esters, 40
sugar complexes by EPR, 188
sugar components of antibiotic substances, 89
sugar field, application of neighboring-group reactions in, 73
sugar hydrazones and osazones, 21–22
sugar mutarotation, phenomenon of, 56
sugar nitrates, 34
sugar "osazones", 50
sugar oxygen rings, formation and cleavage of, 36
sugars and derivatives, analysis of, 65
sugars, branched-chain, 32
sugars, internal anhydrides of, 157–158
sugars, physicochemical properties of, 40
sugars, polysaccharides and glycosides, 145
sugars, preparation, properties, and uses of, 67
sugars reaction, phenylhydrazine, 146–147
sugars, reactions of with ammonia and amines, 62–63
sugars, selenium-containing, 140
sulfate half-esters, 50
sulfonic esters of carbohydrates, 22, 27, 59
surface glycolipids of mycobacteria, 135
(−)-swainsonine, 188
synthesis and applications of polysaccharides, 73
synthesis and polymerization of anhydro sugars, 101
synthesis and reactions of the gulono-1,4-lactones and derivatives, 99
synthesis of D-glucuronic acid, 22
synthesis of glycopeptides, 132

synthetic cardenolides/cardiac glycosides, 52
tabulated carbon-13 data for monosaccharides, 104
Tamiflu®, 168
tautomeric composition of reducing sugars in solution, 127
tautomeric interconvention of sugars, 127
techniques for polysaccharide structure-determination, 83
technology of polymeric carbohydrates, 160
teichoic acids, 52
tetraacetate, lead, 38
thermal decomposition of sugars, 123
thermal degradation of starch, 54
thiamine biosynthesis, 138
thio and seleno sugars, 20
N-thiocarbonyl derivatives of carbohydrates, 146
thiocyanates of monosaccharides, 112
thioglycosides as glycosyl donors, 137
thio sugars, 46
Thompson, Alva, portrait, 49
three-dimensional carbohydrate structures and protein binding, 158
Tipson, Robert Stuart, portrait, 133
Todd, Alexander R. (Lord), portrait, 148
traditional chromatographic methods, 119
trehaloses, 46
trehalozin synthesis, 184
trichloroacetimidate methodology, 131, 170–171
trityl ethers, 21–22
Umezawa, Hamao, portrait, 126
Umezawa, Sumio, portrait, 156
unique group of natural cryptands, 119
"unnatural" monosaccharide enantiomers, 103
unsaturated sugars, 59, 157
use of boric acid for the determination of the configuration of carbohydrates, 22
utilization of sucrose, 22
utilization of sugars by yeasts, 101
Vargha, Laszlo, portrait, 71

Articles (*Continued*)
  vibrational spectra of carbohydrates, 112
  vitamin C (L-ascorbic acid), practical synthesis of, 96
Whistler, Roy Lester, portrait, 182
Wittig reaction in carbohydrate chemistry, 67
Wolfrom, Melville Lawrence, portrait, 66
wood, non-cellulosic constituents of, 30
wood, polysaccharides accompanying cellulose in, 50
wood saccharification, 22
xanthan, 93
X-ray crystallographers, 110–111
X-ray crystallographic and electron-diffraction methods, 84, 104
X-ray crystallography, applications of, 134
X-ray crystal-structure analysis, 62
X-ray diffraction analysis, 138, 202
X-ray structure of polysaccharides, 54
xylan, 23
yeasts, utilization of sugars by, 80
Zamojski, Aleksander, portrait, 166
Zemplén, Geza, portrait, 39
zeolites, 175
zone electrophoresis, 34
Zorbach, William Werner, portrait, 68